Springer Theses

Recognizing Outstanding Ph.D. Research

Aims and Scope

The series "Springer Theses" brings together a selection of the very best Ph.D. theses from around the world and across the physical sciences. Nominated and endorsed by two recognized specialists, each published volume has been selected for its scientific excellence and the high impact of its contents for the pertinent field of research. For greater accessibility to non-specialists, the published versions include an extended introduction, as well as a foreword by the student's supervisor explaining the special relevance of the work for the field. As a whole, the series will provide a valuable resource both for newcomers to the research fields described, and for other scientists seeking detailed background information on special questions. Finally, it provides an accredited documentation of the valuable contributions made by today's younger generation of scientists.

Theses are accepted into the series by invited nomination only and must fulfill all of the following criteria

- They must be written in good English.
- The topic should fall within the confines of Chemistry, Physics, Earth Sciences, Engineering and related interdisciplinary fields such as Materials, Nanoscience, Chemical Engineering, Complex Systems and Biophysics.
- The work reported in the thesis must represent a significant scientific advance.
- If the thesis includes previously published material, permission to reproduce this must be gained from the respective copyright holder.
- They must have been examined and passed during the 12 months prior to nomination.
- Each thesis should include a foreword by the supervisor outlining the significance of its content.
- The theses should have a clearly defined structure including an introduction accessible to scientists not expert in that particular field.

More information about this series at http://www.springer.com/series/8790

Alessandro Franci

Unified Lagrangian Formulation for Fluid and Solid Mechanics, Fluid-Structure Interaction and Coupled Thermal Problems Using the PFEM

Doctoral Thesis accepted by
Universitat Politècnica de Catalunya, Barcelona, Spain

 Springer

Author
Dr. Alessandro Franci
Universitat Politècnica de Catalunya
Barcelona
Spain

and

CIMNE
Barcelona
Spain

Supervisors
Prof. Eugenio Oñate
Barcelona
Spain

Dr. Josep Maria Carbonell
Barcelona
Spain

ISSN 2190-5053   ISSN 2190-5061   (electronic)
Springer Theses
ISBN 978-3-319-83341-5   ISBN 978-3-319-45662-1   (eBook)
DOI 10.1007/978-3-319-45662-1

Printed on acid-free paper

This Springer imprint is published by Springer Nature
The registered company is Springer International Publishing AG
The registered company address is: Gewerbestrasse 11, 6330 Cham, Switzerland

Dai diamanti non nasce niente,
dal letame nascono i fior.

Nothing grows out of precious diamonds,
Out of dung the flowers do grow.

Fabrizio De André

To my mother and father

Supervisor's Foreword

It is a big pleasure to write a foreword for Dr. Alessandro Franci's thesis. The publication in the Springer Thesis program represents a further recognition of the quality of thesis, which has received the PIONER prize and the SEMNI award conferred in 2015 by the Catalan association CERCA and the Spanish Society of Numerical Methods in Engineering (www.semni.com), respectively.

The objective of Dr. Franci's thesis was the development and experimental validation of a new particle-based computational method termed particle finite element method (PFEM) for the solution of practical fluid–structure interaction (FSI) problems.

The PFEM formulation proposed in the thesis represents an extremely powerful numerical tool that can be used for a wide range of engineering problems. In particular, the PFEM is very adequate to reproduce bulk forming processes, such as the forging and extrusion of metal pieces, among others. The numerical modeling of these industrial problems is useful for the optimization of the manufacturing process and the reduction of the defects in the final products.

The thesis includes many practical engineering applications (not only of forming processes), such as the collapse of a water column against a deformable membrane and the melting of ice block in a tank filled with hot water, among others.

The last chapter of the thesis has been devoted to an industrial application solved with the proposed computational method. The numerical results showed in this chapter were obtained by Dr. Franci as part of an international project sponsored by a Japanese company carried out in the three-month period during June–September of 2014.

The object of the project was to simulate two hypothetical scenarios during a nuclear core melt situation, one of the most severe accident in a nuclear power plant. This kind of analysis belongs to those problems that are difficult to solve with traditional strategies or with laboratory tests and also their numerical simulation is extremely complex. Despite that, with the methodology developed by Dr. Franci in his thesis fine and accurate simulations of this phase-change problem were obtained.

The success of this project was another proof of the applicability of the numerical strategy proposed in the thesis of Dr. Franci for solving industrial problems.

The numerical formulation has been well received also in the scientific community. I highlight that six publications in international peer-reviewed journals have already resulted from Dr. Franci's thesis.

The thesis outcomes leave many open lines of research and future developments. In particular, there is the high interest to couple the PFEM technique developed in the thesis with the discrete element method (DEM) and to extend the proposed formulation to large scale engineering problems.

Knowing the motivation, dedication and capabilities of Dr. Franci, I look forward to seeing him leading these new projects in the next coming years.

Barcelona, Spain   Prof. Eugenio Oñate
May 2016

Parts of this thesis have been published in the following articles:

E. Oñate and A. Franci and J.M. Carbonell, *Lagrangian formulation for finite element analysis of quasi-incompressible fluids with reduced mass losses*, International Journal for Numerical Methods in Fluids, 74 (10), 699–731, 2014;

E. Oñate and A. Franci and J.M. Carbonell, *A particle finite element method (PFEM) for coupled thermal analysis of quasi and fully incompressible flows and fluid-structure interaction problems*, Numerical Simulations of Coupled Problems in Engineering. S.R. Idelsohn (Ed.), 33, 129–164, 2014;

E. Oñate and A. Franci and J.M. Carbonell, *A particle finite element method for analysis of industrial forming processes*, Computational Mechanics, 54, 85–107, 2014;

A. Franci and E. Oñate and J.M. Carbonell, *On the effect of the bulk tangent matrix in partitioned solution schemes for nearly incompressible fluids*, International Journal for Numerical Methods in Engineering, 102, 257–277, 2015;

A. Franci and E. Oñate and J.M. Carbonell, *Unified formulation for solid and fluid mechanics and FSI problems*, Computer Methods in Applied Mechanics and Engineering, 298, 520–547, 2016;

A. Franci and E. Oñate and J.M. Carbonell, *Velocity-based formulations for standard and quasi-incompressible hypoelastic-plastic solids*, International Journal for Numerical Methods in Engineering, 10.1002/nme.5205, 2016.

Acknowledgements

I would like to thank all the people that in some way have helped me to complete this doctoral thesis.

I would like to express my gratitude to Prof. Eugenio Oñate for many reasons. Thanks to him, four years ago I could come to Barcelona and start my Ph.D. at CIMNE. He gave me the balanced doses of freedom and pressure and he encouraged me to test innovative technologies. I would like thank Prof. Oñate also for giving me the opportunity and the material support to spend a three-month period at the College of Engineering of Swansea.

Special thanks go to Dr. Josep Maria Carbonell. Without his support I could not realize this work. He devoted me innumerable hours of his time and he gave me technical and non-technical suggestions that helped me to make the right choices during my Ph.D.

I would like to thank Prof. Javier Bonet for giving me the possibility to work with his group during my stay at Swansea. Special thanks go to Prof. Antonio Gil y Dr. Aurelio Arranz for their very kind treatment, for improving my knowledge of the Immersed boundary potential method and for the useful discussions about FSI strategies. I really hope that this is just the first step for a future fruitful collaboration.

I also acknowledge the Agencia de Gestió d'Ajuts Universitaris i de Recerca (AGAUR) for the financial support.

I would like to thank also Riccardo, Antonia, Jordi, Pablo, Lorenzo and Stefano for their contributions to this thesis.

This thesis is dedicated to my parents who still represent to me a fundamental and unique support.

Finally, I would like to thank Lucía simply for staying by my side during these years. Her smile has been the gasoline I used for dealing with all this work.

Contents

Abbreviations

AL	Augmented Lagrangian
ASGS	Algebraic SubGrid Scale
FCM	Fuel Containing Material
FEM	Finite Element Method
FIC	Finite Increment Calculus
FSI	Fluid–Structure Interaction
GLS	Galerkin Least Squares
IBM	Immersed Boundary Method
IFEM	Immersed Finite Element Method
ISPM	Immersed Structural Potential Method
LBB	Ladyzenskaya–Babuzka–Brezzi
LFCM	Lava-like Fuel Containing Material
NPP	Nuclear Power Plant
OSS	Orthogonal SubScale
PFEM	Particle Finite Element Method
SPH	Smooth Particle Hydrodynamics
SUPG	Streamline Upwind Petrov Galerkin
TL	Total Lagrangian
UL	Updated Lagrangian
V	Velocity
VMS	Variational Multi-Scale
VP	Velocity-Pressure
VPS	Velocity-Pressure-Stabilized

Chapter 1
Introduction

The objective of this work is to develop a unified formulation for the solution of fluid and solid mechanics, Fluid-Structure Interaction (FSI) and thermal coupled problems and to prove its efficacy by solving both academic and industrial problems.

Due to their complexity, generally FSI problems cannot be solved with the traditional engineering methodology and numerical methods represent the best alternative to the expansive laboratory tests and even, in same cases, the unique possibility to face them.

FSI problems involve a large number of physical phenomena characterizing many fields of engineering, technology and also biology. An example of problem of interest in civil engineering is the safety study of civil constructions to water-induced hazards. These constructions include: buildings, bridges, harbors, dams, dykes, breakwaters, and similar infrastructures in water hazard scenarios such as flooding, large sea waves, tsunamis and water spills due to the collapse of dams, dykes and reservoirs, among others. Also industrial engineering is full of example of complex FSI problems. For example, this is the case of many manufacturing processes in which complicated thermo-coupled interactions occur between different materials at different phases. This list could be further extended considering other branches of engineering, from aeronautics to mechanical and naval engineering.

The numerical method developed in this work is designed for solving a big part of the mentioned situations.

From the theoretical point of view, the aim is to analyze the continuum in a unified manner trying to reduce at minimum the differences between the analysis of fluids and solids. For this purpose the numerical model has been designed in order to meet the specific requirements of solid and fluid mechanics and their approximation with the FEM, but without limiting excessively the capability of the model. In fact the computational method should be capable to deal with critical problems such as those involving elastic-plastic solids, quasi-incompressible materials, free-surface fluids and phase change.

© Springer International Publishing AG 2017
A. Franci, *Unified Lagrangian Formulation for Fluid and Solid Mechanics,*
Fluid-Structure Interaction and Coupled Thermal Problems Using the PFEM,
Springer Theses, DOI 10.1007/978-3-319-45662-1_1

Following these considerations, the computational model has been designed according to a stabilized Velocity–Pressure formulation. The numerical method has been applied for solving hypoelasto-plastic, compressible and quasi-incompressible solids and quasi-incompressible Newtonian fluids. The algorithm for the FSI problems has been inspired by the analogous unified strategy presented in [1]. For the fluid phase, the Particle Finite Element Method (PFEM) [2] has been used, while for the solid the classical Finite Element Method (FEM) [3] is adopted.

The Unified formulation is based on a stabilized Velocity–Pressure Lagrangian procedure. Each time step increment is solved using a two-step Gauss–Seidel scheme: first the linear momentum equations are solved for the velocity increments, next the continuity equation is solved for the pressure in the updated configuration.

Linear shape functions are used for both the velocity and the pressure fields. In order to deal with the incompressibility of the materials, the formulation has been stabilized using an updated version of the Finite Calculus (FIC) method [4]. The procedure has been derived for quasi-incompressible Newtonian fluids. In this work, the FIC stabilization procedure has been extended also to the analysis of quasi-incompressible hypoelastic solids [5].

Specific attention has been given to the study of free surface flow problems. In particular, the mass preservation feature of the PFEM-FIC stabilized procedure has been deeply studied with the help of several numerical examples. Furthermore, the conditioning of the problem has been analyzed in detail describing the effect of the bulk modulus on the numerical scheme. A strategy based on the use of a pseudo bulk modulus for improving the conditioning of the linear system is also presented.

The Unified formulation has been validated by comparing its numerical results to experimental tests and other numerical solutions for fluid and solid mechanics, and FSI problems. The convergence of the scheme has been also analyzed for most of the problems presented.

The Unified formulation has been coupled with the heat transfer problem using a staggered scheme. A simple algorithm for simulating phase change problems is also described. The numerical solution of several FSI problems involving the temperature is given.

The thermal coupled scheme has been used successfully for the solution of an industrial problem. The objective of study was to analyze the damage of a nuclear power plant pressure vessel induced by a high viscous fluid at high temperature, the corium. The numerical study of this industrial problem has been included in the thesis.

The whole formulation has been implemented in a C++ code.

1.1 State of the Art

In this section, an overview of the numerical methods used for simulating a free surface fluid flow interacting with deformable solids is given. For the sake of clarity, the section is divided in three parts representing the main topics raised by this work. First the Eulerian and Lagrangian approaches for free surface fluid dynamics problems are presented. Then an overview of the stabilization for incompressible, or quasi-incompressible, material is given. Finally, the principal algorithms for solving FSI problems are described.

1.1.1 Eulerian and Lagrangian Approaches for Free Surface Flow Analysis

Consider the description of the motion of a general continuum represented in Fig. 1.1. The domain Ω_0 represents the body at the initial state at time $t = t_0$ while the domain Ω represents the same body at time $t = t_n$ after deformation. The domain Ω_0 is called *initial configuration*, whereas Ω is the *current*, or *deformed*, *configuration*. In order to describe the kinematics and the deformation of the body, the *reference configuration* has to be defined because the motion is defined with respect to this configuration.

In solid mechanics, the stresses generally depend on the history of deformation and the undeformed configuration must be specified in order to define the strains. Due to the history dependence, Lagrangian descriptions are prevalent in solid mechanics. In Newtonian fluids however, the stresses do not depend on the history and it is often unnecessary to describe the motion with respect to a reference configuration. For this reason an Eulerian description represents the most reasonable choice. Furthermore, in problems where the fluid contours are fixed, Eulerian meshes are generally preferred to the Lagrangian ones. This is because Eulerian grids are fixed and they do not deform according to the fluid motion, as shown in Fig. 1.2.

Conversely, in the Lagrangian description, the mesh nodes coincide with the fluid particles and the discretization moves and deforms as the fluid flow (Fig. 1.3).

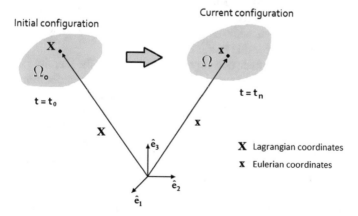

Fig. 1.1 Description of the motion of a general continuum body

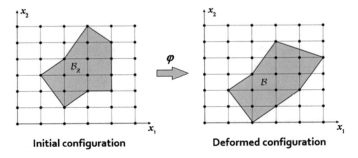

Fig. 1.2 Motion description using an Eulerian mesh

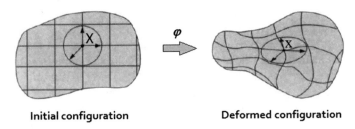

Fig. 1.3 Motion description using a Lagrangian mesh

Consequently, on the one hand the non-linear convective term disappears from the problem and on the other hand the mesh undergoes large distortions and it requires to be regenerated.

In the analysis of free surface flows, the detection of free surface contours represents a crucial task. Its position is unknown a priori and it has to be determined at each time increment in order to solve properly the boundary value problem. For these problems, the Lagrangian description may be preferred to the Eulerian one. In fact, with a Lagrangian approach the free surface is automatically detected by the position of the mesh nodes, while an Eulerian approach requires the implementation of a specific technology for this task.

Several strategies have been developed and presented in the literature for tracking the free surface in an Eulerian framework. One of the earliest contributions was given by the so called *marker and cell* method [6]. In this approach a set of marker particles that move according the flow are used to detect which regions are occupied by the fluid and which not. An evolution of this technique is the *volume of fluid* method [7]. In this case the free surface boundary is detected using a scalar function that assumes the unit value in the fluid cells and the value zero in those ones with no fluid. The cells with an intermedium value are the ones that contain the free surface. Another possibility is the *level set* method [8]. This technique is used in various fields, not only in continuum mechanics, and it allows for detecting shapes or surfaces on a fixed grid without making any parametrization of them. For this reason, this procedure has been also used for matching the free surface contour on an Eulerian mesh [9]. A similar idea was used in [10] where the position of the free surface is detected using a cloud of Lagrangian particles moving over an Eulerian mesh.

The free surface flows can be solved also using an hybrid Eulerian-Lagrangian technique. This is the so termed *Arbitrary Lagrangian-Eulerian (ALE)* approach [11]. The aim of this method is to exploit the best features of the Eulerian and the Lagrangian procedures and to combine them. The mesh nodes can arbitrary be fixed or can move with the fluid [12]. Generally, far from the moving boundaries a fixed Eulerian grid is used, while near to the interface the mesh moves according to the motion of the boundary [13]. However if the boundary motion is large or unpredictable, also in the ALE methods the grid may suffer large distortions and may require a proper remeshing procedure [14].

In purely Lagrangian approaches, the mesh needs to be regenerated whenever a threshold limit for the distortion is exceeded. This is the basis of a particular class of Lagrangian finite element formulation called the *Particle Finite Element Method (PFEM)*. The method was initially developed by the group of professors Idelsohn and Oñate [15, 16]. The PFEM treats the mesh nodes of the domain as particles which can freely move and even separate from the rest of the fluid domain representing, for instance, the effect of water drops. A mesh connects the nodes discretizing the domain where the governing equations are solved using a classical finite element method. These features make the PFEM the ideal numerical procedure to model and simulate free surface flows. In the last years, many scientific publications have shown the efficacy of the PFEM for solving free surface flow problems, see among others [17–19]. The PFEM has also successfully been tested in other kind of problems, such as fluid mechanics including thermal convection-diffusion [20–22], multi-fluids [23, 24], granular materials [25], bed erosion [26], FSI [27, 28] and excavation [29].

Meshfree methods are other class of Lagrangian techniques. In this strategy the remeshing is not required because the governing equations are solved over a set of nodes without referring to a mesh. One of the first meshfree techniques is the *Smooth Particle Hydrodynamics (SPH)* method. This method was introduced independently by Gingold and Monaghan [30] and Lucy [31] for the simulation of astrophysical problems such as fission of stars. SPH is a particle-based Lagrangian technique where discrete smoothed particles are used to compute approximate values of needed physical quantities and their spatial derivatives. The particles have assigned a characteristic distance, called 'smoothing length', over which their properties are "smoothed" by a kernel function. A typical drawback of the SPH method is that it is hard to reproduce accurately the incompressibility of the materials. The SPH technique has been used successfully for solving fluid-structure interaction problems [32].

1.1.2 Stabilization Techniques

A FEM-based procedure may require to be stabilized when incompressible, or quasi-incompressible, materials are analyzed. In the FEM solution of the Navier Stokes problem numerical instabilities may arise from two sources. The first one is due to the presence of the convective term in the linear momentum equations. This term introduces a non-linearity in the equations and it needs a proper stabilization for solving high Reynolds number flows with the FEM [12]. Furthermore, the orders of interpolation of pressure and velocity fields cannot be chosen freely but they have to satisfy the so called *Ladyzenskaya–Babuzka–Brezzi* (LBB), or *inf–sup*, condition [33]. If the orders of interpolation of the unknown fields do not satisfy this restriction, a stabilization technique is required in order to avoid numerical instabilities, as the spurious oscillations of the pressure field.

It is well known that the weak form generated by Galerkin approximation leads to a less diffusive solution than the strong problem. So the main idea of many stabilization techniques consists of adding an artificial diffusion to the problem. The first attempt was made by Von Neumann and Richtmyer [34]. Their solution adds an artificial diffusion to the strong form of the problem. However, this technique introduces an excessive dissipation to the problem because the diffusivity is added in every direction. An important evolution of this approach was the *Streamline Upwind Petrov Galerkin (SUPG)* [35]. In this approach, the artificial diffusion is added by means of the test functions and only on the direction of the streamlines. Furthermore, this is performed in a consistent way: the stabilization terms vanish when the solution is reached. An extension of the SUPG method is the *Galerking Least-Squares (GLS)* method [36]. In the GLS method the stabilization terms are applied not only to the convective term but to all the terms of the equation. The *Variational Multi-Scale (VMS)* methods [37] split the problem variables in a large-scale and a subscale terms. The large-scale terms represent the part of the solution that can be captured by the finite element mesh, while the subscale part consists of an approximate solution that has to be added to the large-scale term in order to obtain the correct solution. This idea represents the basis of other stabilization methods. Among these, the most largely used are the *Algebraic Subgrid Scale Formulation (ASGS)* and the *Orthogonal Subscales (OSS)* methods, respectively introduced in [38, 39]. Another efficient stabilization technique is the *Finite Calculus* (also termed *Finite Increment Calculus*) *(FIC)* approach (see among others [40–43]). This method has some analogies with the SUPG technique (for a comparison between these methods see [44]). The FIC approach is based on expressing the equations of balance of mass and momentum in a space-time domain of finite size and retaining higher order terms in the Taylor series expansion typically used for expressing the change in the transported variables within the balance domain. In addition to the standard terms of infinitesimal theory, the FIC form of the momentum and mass balance equations contains derivatives of the classical differential equations in mechanics multiplied by characteristic distances in space and time. In this work an updated version of the FIC method has been derived and tested.

In solid mechanics a stabilization procedure may be required for the solution of problems involving incompressible, or quasi-incompressible solids. Situations of this type are common in forming processes or in the analysis of rubber-type materials. Many of the mentioned stabilization procedures for fluid dynamics have been also used also for solid dynamics. For example, the VMS method has been applied in quasi-incompressible solid mechanics in [45, 46], the OSS in [47]. In [48, 49] a stabilized multi-field Petrov–Galerkin procedure is used. Finally, the application of the FIC in solid mechanics is reported [50, 51].

1.1.3 Algorithms for FSI Problems

Many approaches have been developed for solving FSI problems. Typically the computational techniques for FSI are distinguished in *monolithic* and *staggered*, or *partitioned*, approaches. In a monolithic approach fluid and solid domains are solved in a single system of equations (see among others [52], or [53]). In this technique the flow of information between the solid and the fluid parts is implicitly performed by the procedure. On the contrary, in staggered schemes the fluid and the solid dynamics are solved separately and boundary conditions are transferred from a domain to the other at the interface. From the algebraic point of view, the solution is achieved by solving two different linear systems which are coupled by means of the boundary conditions defined along the interface. Within partitioned schemes, a further classification can be done depending on the level of coupling between the fluid and the solid dynamics. In *weakly coupled* segregated methods the transfer of information at the interface is performed only once for each time step, (for example see [54]), whereas in *strongly coupled* schemes this operation is performed within a convergence iterative loop (as a reference see [55]). Clearly, a weakly coupled scheme has a lower computational cost but it can be used only when the interaction between the solid and the fluid domains is not strong or complex. Otherwise the algorithm may not find the correct numerical solution of the problem.

Partitioned methods allow the reutilization of existing solvers. Furthermore, the solid and the fluid solvers can be updated independently. Staggered schemes lead to smaller and better conditioned linear systems than the ones obtained with monolithic approaches. On the other hand, monolithic strategies are generally more stable than staggered schemes, and they lead to a more accurate solution of FSI problems, because a stronger coupling is ensured.

Another general classification of the FSI algorithms is based upon the treatment of meshes. In *conforming mesh* methods, the fluid and the solid meshes must have in common the nodes along the interface in order to allow the transfer of information. Consequently, if the position of the interface nodes changes in a material domain, also the other domain must modify its grid in order to guarantee the conformity of the two meshes along the interface. On the contrary, in *non-conforming mesh* methods the interface and the related conditions are treated as constraints imposed on the model equations so that the fluid and solid equations can be solved independently from each other with their respective grids [56]. This represents an important advantage because, typically, the mesh used for the fluid has an average size lower than the one used for the solid and so it is not necessary to refine the solid finite element grid near to the interface. However, non-conforming mesh algorithms are more complex to implement and it is not trivial to guarantee their robustness.

In the so-called *Immersed Boundary Method* (*IBM*), the fluid is solved using an Eulerian grid and the solids are immersed on top of this mesh [57, 58]. The interaction is ensured by penalizing the Navier–Stokes equations with the momentum forcing sources of the immersed structures. An evolution of the IBM is the *Immersed Structural Potential Method* (*ISPM*) where the structure is modeled as a potential energy functional solved over a cloud of integration points that move within the fixed fluid mesh [58, 59]. Also in the *Immersed Finite Element Method* (*IFEM*) [60, 61] the structure acts as a momentum forcing source for the fluid governing equations, but in this case a Lagrangian mesh is employed for the solid domains.

1.2 Numerical Model

The aim of this work is to derive a finite element formulation capable to solve, through a unique set of equations and unknown variables, the mechanics of a general continuum. The term 'general continuum' refers to a domain that may include compressible and quasi-incompressible solids, fluids or both interacting together. For this reason, the formulation is termed *Unified*. The Unified formulation is based on a mixed Velocity–Pressure Stabilized procedure and it has been implemented in a sequential C++ code.

1.2.1 Reasons

There are many reasons for undertaking the above objective.

The first advantage of the Unified formulation is that it allows for solving fluid and solid dynamics problems by implementing and using a single calculation code.

Furthermore, if solids and fluids are solved via the same scheme, it is simpler to implement the solver for FSI problems because it is not required neither changing the variables, neither implementing the transfer of transmission conditions through the interface. With this formulation solids and fluids represent regions of the same continuum and they differ only by the specific values of the material parameters. As a consequence, the FSI solver requires a small computational effort and it can be implemented by introducing just a few specific functions. This will be explained in detail in the section dedicated to FSI problems.

Additionally, with the Unified formulation the most natural approach for solving FSI problem is the monolithic one. This brings in the further advantage that the coupling is ensured strongly and an iteration loop is not required, as for staggered procedures.

Finally, using the same set of unknowns for the fluid and the solid domains reduces the ill-conditioning of the FSI solver, because the solution system does not include variables of different units of measure.

1.2.2 Essential Features

The Unified formulation has been designed as a compromise between a fluid and a solid formulation in order to make it capable to satisfy the requirements of each physics and finite element approximations. In other words, in order to solve adequately fluids and solids with the same formulation, the solution procedure must take into account the constraints that both models impose. In this part, the main considerations made for designing the numerical formulation developed in this thesis are recalled.

The choice of the unknown variables is generally driven by the specific constitutive law used for the material description. For Newtonian fluids, the Cauchy stress tensor is related directly to the deformation rate tensor. This explains why the velocities are the most used unknowns in fluid mechanics. Conversely, velocity-based schemes are not largely employed in solid mechanics, where generally, displacement-based approaches are preferred. The main reason is that velocity formulations require a time integration procedure also for computing the stresses of an elastic material, as for the inelastic ones. However, there also exist solid constitutive laws expressed in rate form, as for hypoelastic models, for which a velocity formulation can be useful.

For the analysis of incompressible, or quasi-incompressible, materials a mixed formulation is required in order to overcome the associated numerical instabilities, in solids as in fluids. In fluids, the most popular combination of unknown variables is the Velocity–Pressure scheme. In solid mechanics, several mixed approaches have been presented, as the displacement-stress [62], the displacement-strain [63], the displacement-deformation gradient [48], the displacement-deformation gradient–jacobian [49], the displacement-pressure [47] and the Velocity–Pressure formulation [64]. The combination of unknowns is selected depending on the desired accuracy for the stress and strain fields and the associated computational cost. Among the mentioned mixed approaches, the displacement-pressure and the Velocity–Pressure formulations lead to the lowest computational cost, because the number of unknowns is smaller. However the Velocity–Pressure formulation has the further advantage that is the canonical scheme for fluid mechanics. Hence, according to these considerations, the Unified formulation has been based on a mixed Velocity–Pressure scheme.

Another key decision in the design of the Unified formulation concerns the choice of the description framework. Solids are typically solved using a (total or updated) Lagrangian description, while for fluids, due to the large deformations they undertake, the Eulerian framework is generally preferred. However, for free surface flows, as the ones treated in this work, a Lagrangian description has the important advantage that the free surface boundaries are detected automatically, because the particles of the fluid coincide with the nodes of the mesh. The price for this is that a remeshing procedure is required in order to avoid the excessive distortion of the mesh. According to these considerations, the Unified formulation has been implemented using an updated Lagrangian description.

In this work, the Particle Finite Element Method (PFEM) [2] has been used for solving the fluid domain only. The PFEM is a Langrangian numerical technique that treats the mesh nodes as moving points which can freely move and even separate from the domain to which they are initially attached representing, for instance, the effect of water drops. The PFEM is based on a remeshing procedure that efficiently combines the Delaunay tessellation and the Alpha Shape method [65]. In order to reduce the computational cost associated to the remeshing step, all the unknowns are stored in the nodes and linear shape functions are used for the finite element interpolation. Conversely, the solid domain keeps a fixed mesh during the whole analysis. The reason for this is that in non-linear solid mechanics it is required to preserve not only the nodal unknowns but also elemental information as the stress state of the previous time step. A remeshing implies the elimination of the previous discretization and the creation of new elements. In order to keep the elemental information in the remeshing step, a projection procedure from the elements of the previous mesh to its nodes, and from these to the elements of the new mesh is required. These operations may introduce an interpolation error in the scheme that affects the solution of solid mechanics problems. In the analysis of Newtonian fluids all the information can be stored in the nodes, so the remeshing does not affect the accuracy of the scheme. For these reasons, in this work the PFEM is used only for the fluid, while solids are solved via the classical FEM.

In order to improve the conditioning of the solution system, a Gauss–Seidel segregated scheme is used. This means that the problem is solved iteratively and separating the unknown variables, the velocities and the pressure. The solution method consists of two steps. In the first one the linear momentum equations are solved for the velocity increments and considering the pressure of the previous non-linear iteration in the residual term. In the second step, the continuity equation is solved for the pressure in the updated configuration using the velocities computed at the first step. This two-step algorithm leads to a smaller and better conditioned linear system than using a monolithic approach. A crucial point of this scheme is the derivation of the tangent matrix of the linear momentum equations. The solution for the velocity increments is obtained by condensing the pressure. According to this procedure, the tangent matrix for the linear momentum equations of this mixed Velocity–Pressure formulation does not differ from the one of a velocity formulation. Furthermore, the derivation of the tangent matrix for the velocity formulation is easier, as it does not involve the pressure. For these reasons, the first step towards the unified scheme is the derivation of the Velocity formulation. The mixed Velocity–Pressure formulation will be derived by exploiting the linearization performed for the Velocity formulation. In [66] another procedure for the derivation of the exact tangent matrix for an Updated Lagrangian scheme is described.

The condensation of the pressure in the tangent matrix of the linear momentum equations may induce the ill-conditioning of the linear system for the analysis of quasi-incompressible materials. This ill-conditioning emanates from the volumetric counterpart of the tangent matrix that is governed by the bulk modulus, that typically has high values that may compromise the conditioning of the linear system. In this work, a thorough study of the conditioning drawbacks associated to this scheme has been performed and a useful and easy to implement technique to overcome these inconveniences has been tested and validated.

The numerical method proposed in this work is an improvement versus other numerical schemes for quasi-incompressible materials where an arbitrarily defined pseudo-bulk modulus was used in both the linear momentum and the mass conservation equations [64, 67, 68].

The method can be also related to the so-called Augmented Lagrangian (AL) procedures for solving the Navier–Stokes equations for incompressible [69–73] and weakly compressible [74, 75] flows.

In order to deal with incompressible (or quasi-incompressible) materials a mixed formulation is required. In this work both the velocities and the pressure are interpolated using linear shape functions. This combination does not fulfil the *inf–sup* condition [33] and the problem needs to be stabilized. The required stabilization is ensured using an updated version of the FIC technique [4] applied to the mixed Velocity–Pressure formulation. The FIC stabilization method has a small intrusivity because its terms only affect the continuity equation. In fact for solving the linear momentum equations the same scheme as for the (not stabilized) mixed Velocity–Pressure formulation is used. It will be shown that the PFEM-FIC stabilized procedure guarantees a good accuracy for the mass conservation of the free surface flows. The stabilization procedure is derived for quasi-incompressible fluids, but it can be easily extended to quasi-incompressible hypoelastic solids.

The derivation of the Unified stabilized formulation is carried on trying to maintain as much as possible the generality of the scheme. For this reason the constitutive relations are introduced only as the last step. The Velocity (V) and the mixed Velocity–Pressure (VP) finite elements are derived first for a general material and then particularized for specific constitutive laws. For quasi-incompressible materials only the mixed Velocity–Pressure Stabilized (VPS) finite element can be used while for compressible materials the Velocity and the standard (not stabilized) mixed Velocity–Pressure formulations are both suitable. In particular, for compressible solids, the hypoelastic model has been chosen.

The FSI problem is solved in a monolithic way, hence solids and fluids are solved by the same linear system. For the solid domain it is possible to choose which formulation to use. Depending on the problem, one may chose the V, the VP or the VPS element.

The Unified formulation for FSI problems can be easily coupled with the heat transfer problem in order to solve coupled thermal mechanical problem. The coupling is performed via a staggered scheme. The effect of the heat is taken into account by considering the material properties depending on the temperature and including in the strain tensor the deformations induced by the temperature. With the PFEM Unified formulation also phase change problems can be modeled.

The Unified formulation with thermal coupling has been used for the analysis of an industrial application. The study concerned the analysis of the damages caused by the dropping of a volume of corium at high temperature on the structure of the pressure vessel in a Nuclear Power Plant (NPP).

1.2.3   Outline

This text is split in the following chapters.

In the next chapter two velocity-based FEMs are derived for a general compressible material. In the first section the Velocity formulation is derived and the incremental solution scheme is given. Then, the standard mixed Velocity–Pressure formulation is obtained as an extension of the Velocity formulation. The constitutive laws are not introduced for any scheme. In the third section the formulations are adapted for hypoelastic-plastic compressible solids. The chapter ends with several validation examples for non-linear solid mechanics.

In Chap. 3, the FIC stabilization strategy is introduced in the mixed stabilized VP scheme. This scheme is adapted for quasi-incompressible Newtonian fluids and hypoelastic solids, in this order. Then the free surface problem is studied in detail. First the PFEM is described highlighting the advantages and the disadvantages of the method. A useful technique for modeling the slip conditions in Lagrangian flows is also explained. Next the mass conservation feature and the conditioning of the scheme are analyzed in detail. Concerning the former point, it will be shown that the PFEM-FIC formulation guarantees a good preservation of the mass in free surface flows. Regarding the latter point, a practical strategy for improving the conditioning of the linear system is explained and tested. At the end of the chapter several validation examples for quasi-incompressible Newtonian fluids and hypoelastic solids are given.

The FSI solution strategy is described in Chap. 4. The monolithic scheme for coupling the mechanics of fluids and solids is explained in detail. Finally several validation examples of FSI problems are given.

Chapter 5 is dedicated to couple the Unified formulation for FSI with the heat problem. In the first section the heat problem is introduced and discretized using the FEM. Then the coupling strategy is explained and validated with numerical examples. Next, the procedure for modeling phase change problems is described and an explicative numerical example is presented.

Chapter 6 is fully devoted to the industrial application of the Unified stabilized and thermally coupled strategy. The damages of the pressure vessel structure caused by the fall of a volume corium at high temperature are studied using two simplified models where the solid structure melts due to the heat transfer from the corium.

In Chap. 7 the innovative contributions of the work are summarized and the lines of research opened by this thesis are presented.

References

1. S.R. Idelsohn, J. Marti, A. Limache, and E. Oñate. Unified lagrangian formulation for elastic solids and incompressible fluids: Applications to fluid-structure interaction problems via the pfem. *Computer Methods In Applied Mechanics And Engineering*, 197:1762–1776, 2008.
2. E. Oñate, S.R. Idelsohn, F. Del Pin, and R. Aubry. The particle finite element method. an overview. *International Journal for Computational Methods*, 1:267–307, 2004.

3. O.C. Zienkiewicz and R.L. Taylor. *The Finite Element Method. Its Basis and Fundamentals. (6th Ed.).* Elsevier Butterworth-Heinemann, Oxford, 2005.

4. E. Oñate, A. Franci, and J.M. Carbonell. Lagrangian formulation for finite element analysis of quasi-incompressible fluids with reduced mass losses. *International Journal for Numerical Methods in Fluids*, 74 (10):699–731, 2014.

5. A. Franci, E. Oñate, and J. M. Carbonell. Velocity-based formulations for standard and quasi-incompressible hypoelastic-plastic solids. *International Journal for Numerical Methods in Engineering*, doi:10.1002/nme.5205, 2016.

6. F. H. Harlow and J. E. Welch. Numerical calculation of time-dependent viscous incompressible flow of fluid with free surface. *Physics of Fluids*, 8:2182, 1965.

7. C.W. Hirt and B.D. Nichols. Volume of fluid (vof) method for the dynamics of free boundaries. *Computational Physics*, 39:201–225, 1981.

8. S.J. Osher and R.P. Fedkiw. *Level Set Methods and Dynamic Implicit Surfaces.* Springer edition, 2006.

9. R. Rossi, A. Larese, P. Dadvand, and E. Oñate. An efficient edge-based level set finite element method for free surface flow problems. *International Journal for Numerical Methods in Fluids*, 71 (6):687–716, 2013.

10. P. Becker, S.R. Idelsohn, and E. Oñate. A unified monolithic approach for multi-fluid flows and fluid-structure interaction using the particle finite element method with fixed mesh. *Computational Mechanics*, 61:1–14, 2015.

11. O.C. Zienkiewicz, R.L. Taylor, and P. Nithiarasu. *The Finite Element Method for Fluid Dynamics, Volume 3 (6th Ed.).* Elsiever, Oxford, 2005.

12. J. Donea and A. Huerta. *Finite Element Methods for Flow Problems.* Wiley, 2003.

13. M. Chiumenti, M. Cervera, C. Agelet de Saracibar, and N Dialami. Numerical modeling of friction stir welding processes. *Computer methods in applied mechanics and engineering*, 254:353–369, 2013.

14. S. Tanaka and K. Kashiyama. Ale finite element method for fsi problems with free surface using mesh re-generation method based on background mesh. *International Journal of Computational Fluid Dynamics*, 20:229–236, 2006.

15. S.R. Idelsohn, E. Oñate, and F. Del Pin. The particle finite element method: a powerful tool to solve incompressible flows with free-surfaces and breaking waves. *International Journal for Numerical Methods in Engineering*, 61:964–989, 2004.

16. PFEM in CIMNE website. www.cimne.com/pfem.

17. A. Larese, R. Rossi, E. Oñate, and S.R. Idelsohn. Validation of the particle finite element method (pfem) for simulation of free surface flows. *International Journal for Computer-Aided Engineering and Software*, 25:385–425, 2008.

18. M. Cremonesi, L. Ferrara, A. Frangi, and U. Perego. A lagrangian finite element approach for the simulation of water-waves induced by landslides. *Computer and Structures*, 89:1086–1093, 2011.

19. B. Tang, J.F. Li, and T.S. Wang. Some improvements on free surface simulation by the particle finite element method. *International Journal for Numerical Methods in Fluids*, 60 (9):1032–1054, 2009.

20. R. Aubry, S. R. Idelsohn, and E. Oñate. Particle finite element method in fluid-mechanics including thermal convection-diffusion. *Computers and Structures*, 83:1459–1475, 2005.

21. E. Oñate, R. Rossi, S.R. Idelsohn, and K. Butler. Melting and spread of polymers in fire with the particle finite element method. *International Journal of Numerical Methods in Engineering*, 81 (8):1046–1072, 2010.

22. E. Oñate, M.A. Celigueta, and S.R. Idelsohn. Modeling bed erosion in free surface flows by th particle finite element method. *Acta Geotechnia*, 1 (4):237–252, 2006.

23. S.R. Idelsohn, J. Marti, P. Becker, and E. Oñate. Analysis of multifluid flows with large time steps using the particle finite element method. *International Journal for Numerical Methods in Engineering*, 75 (9):621–644, 2014.

24. T.S. Dang and G. Meschke. An ale-pfem method for the numerical simuation of two-phase mixture flow. *Computer Methods in Applied Mechanics and Engineering*, 278:599–620, 2014.

25. X. Zhang, K. Krabbenhoft, and D. Sheng. Particle finite element analysis of the granular column collapse problem. *Granular Matter*, 16:609–619, 2014.

26. E. Oñate, S.R. Idelsohn, M.A. Celigueta, and R. Rossi. Advances in the particle finite element method for the analysis of fluid-multibody interaction and bed erosion in free surface flows. *Computer methods in applied mechanics and engineering*, 197 (19–20):1777–1800, 2008.

27. E. Oñate, M.A. Celigueta, S.R. Idelsohn, F. Salazar, and B. Suarez. Possibilities of the particle finite element method for fluid-soil-structure interaction problems. *Computation mechanics*, 48:307–318, 2011.

28. M. Zhu and M. H. Scott. Modeling fluid-structure interaction by the particle finite element method in opensees. *Computers and Structures*, 132:12–21, 2014.

29. J.M. Carbonell, E. Oñate, and B. Suarez. Modeling of ground excavation with the particle finite-element method. *Journal of Engineering Mechanics*, 136:455–463, 2010.

30. R.A. Gingold and J.J. Monaghan. Smoothed particle hydrodynamics - theory and application to non-spherical stars. *Monthly Notices of the Royal Astronomical Society*, 20:375–389, 1977.

31. L. Lucy. A numerical approach to the testing of the fission hypothesis. *Astronomical Journal*, 82:1013–1024, 1977.

32. C. Antoci, M. Gallati, and S. Sibilla. A numerical approach to the testing of the fission hypothesis. *Computers and Structures*, 85:879–890, 2007.

33. F. Brezzi. On the existence, uniqueness and approximation of saddle-point problems arising from lagrange multipliers. *Revue française d'automatique, informatique, recherche opérationnelle. Série rouge. Analyse numérique*, 8(R-2):129–151, 1974.

34. J. Von Neumann and R. D. Richtmyer. A method for the numerical calculation of hydrodynamical shoks. *Journal of Applied Physics International Journal of Computational Fluid Dynamics*, 21:232, 1950.

35. T.J.R. Huges and A.N. Brooks. A multi-dimensional upwind scheme with no crosswind diffusion in: Fem for convection dominated flows. 1979.

36. T.J.R. Hughes, L.P. Franca, and G.M. Hulbert. A new finite element formulation for computational fluid dynamics: Viii. the galerkin/least squares method for advective-diffusive equations. *Computer Methods in Applied Mechanics and Engineering*, 73 (2), 1989.

37. T.J.R. Hughes. Multiscale phenomena: Green's functions, the dirichlet-neumann formulation, subgrid scale models, bubbles and the origins of stabilized methods. *Computer Methods in Applied Mechanics and Engineering*, 127 (1–4):387–401, 1995.

38. R. Codina. A stabilized finite element method for generalized stationary incompressible flows. *Computer Methods in Applied Mechanics and Engineering*, 190 (20–21):2681–2706, 2001.

39. R. Codina. Stabilization of incompressibility and convection through orthogonal sub-scales in finite element methods. *Computer Methods in Applied Mechanics and Engineering*, 190 (13–14):1579–1599, 2000.

40. E. Oñate. A stabilized finite element method for incompressible viscous flows using a finite increment calculus formulation. *Computer Methods in Applied Mechanics and Engineering*, 190 (20–21):355–370, 2000.

41. E. Oñate. Derivation of stabilized equations for advective-diffusive transport and fluid flow problems. *Computer methods in applied mechanics and engineering*, 151:233–267, 1998.

42. E. Oñate and J. García. A finite element method for fluid-structure interaction with surface waves using a finite calculus formulation. *Computer methods in applied mechanics and engineering*, 191:635–660, 2001.

43. E. Oñate, J. Garcí a, S.R. Idelsohn, and F. Del Pin. Fic formulations for finite element analysis of incompressible flows. eulerian, ale and lagrangian approaches. *Computer methods in applied mechanics and engineering*, 195 (23–24):3001–3037, 2006.

44. E. Oñate and M. Manzán. A general procedure for deriving stabilized space-time finite element methods for advective-diffusive problems. *International Journal of Numerical Methods in Fluids*, 31:202–221, 1999.

45. M. Cervera, M. Chiumenti, and R. Codina. Mixed stabilized finite element methods in nonlinear solid mechanics: Part i: Formulation. *Computer Methods In Applied Mechanics And Engineering*, 199:2559–2570, 2010.

46. M. Chiumenti, M. Cervera, and R. Codina. Mixed three-field fe formulation for stress accurate analysis including the incompressible limit. *Computer Methods In Applied Mechanics And Engineering*, 283:1095–1116, 2015.
47. M. Chiumenti, Q. Valverde, C. Agelet De Saracibar, and M.Cervera. A stabilized formulation for incompressible elasticity using linear displacement and pressure interpolations. *Computer Methods In Applied Mechanics And Engineering*, 191:5253–5264, 2002.
48. A.J. Gil, C.H. Lee, J. Bonet, and M. Aguirre. A stabilised petrov-galerkin formulation for linear tetrahedral elements in compressible, nearly incompressible and truly incompressible fast dynamics. *Computer Methods In Applied Mechanics And Engineering*, 276:659–690, 2014.
49. A.J. Gil, C.H. Lee, J. Bonet, and M. Aguirre. Development of a stabilised petrov-galerkin formulation for conservation laws in lagrangian fast solid dynamics. *Computer Methods In Applied Mechanics And Engineering*, 268:40–64, 2014.
50. J. Rojek, E. Oñate, and R.L. Taylor. Cbs-based stabilization in explicit dynamics. *International Journal For Numerical Methods In Engineering*, 66:1547–1568, 2006.
51. E. Oñate, J. Rojek, R.L. Taylor, and O.C. Zienkiewicz. Finite calculus formulation for incompressble solids using linear triangles and tetrahedra. *International Journal For Numerical Methods In Engineering*, 59:1473–1500, 2004.
52. B. Hubner, E. Walhorn, and D.Dinkler. A monolithic approach to fluid-structure interaction using space-time finite elements. *Computer Methods in Applied Mechanics and Engineering*, 193:2087–2104, 2004.
53. C. Michler, S.J. Hulshoff, and E.H. Van Brummelenand R. De Borst. A monolithic approach to fluid-structure interaction. *Computers and Fluids*, 33:839–848, 2004.
54. C.A. Felippa and K.C. Park. Staggered transient analysis procedures for coupled mechanical systems: Formulation. *Computer Methods in Applied Mechanics and Engineering*, 24:61–111, 1980.
55. M. Cremonesi, A. Frangi, and U. Perego. A lagrangian finite element approach for the analysis of fluid-structure interaction problems. *International Journal of Numerical Methods in Engineering*, 84:610–630, 2010.
56. F. Casadei and S. Potapov. Permanent fluid-structure interaction with non-conforming interfaces in fast transient dynamics. *Computer Methods in Applied Mechanics and Engineering*, 193:4157–4194, 2004.
57. C.S. Peskin. Flow patterns around hearth values: a numerical method. *Journal of Computational Physics*, 10:252–271, 1972.
58. A.M. Roma, C.S. Peskin, and M.J. Berger. An adaptive version of the immersed boundary method. *Journal of Computational Physics*, 153 (2):509–534, 1999.
59. A.J. Gil, A. Arranz Carreño, J. Bonet, and O.Hassan. The immersed structural potential method for haemodynamic applications. *Journal of Computational Physics*, 229:8613–8641, 2010.
60. L. Zhang, A. Gerstenberger, X. Wang, and W.K Liu. Immersed finite element method. *Computer Methods In Applied Mechanics And Engineering*, 193:2051–2067, 2004.
61. C. Hesch, A.J. Gil, A. Arranz Carreño, and J. Bonet. On continuum immersed strategies for fluid-structure interaction. *Computer Methods in Applied Mechanics and Engineering*, 247:51–64, 2012.
62. F. Brezzi and M. Fortin. *Mixed And Hybrid Finite Element Methods*. Springer, New York, 1991.
63. M. Cervera, M. Chiumenti, L.Benedetti, and R. Codina. Mixed stabilized finite element methods in nonlinear solid mechanics: Part i: Formulation. *Computer Methods In Applied Mechanics And Engineering*, 285:752–775, 2015.
64. P. Ryzhakov, R. Rossi, S.R. Idelsohn, and E. Oñate. A monolithic lagrangian approach for fluid-structure interaction problems. *Computational Mechanics*, 46:883–899, 2010.
65. H. Edelsbrunner and E.P. Mucke. Three dimensional alpha shapes. *ACM Trans Graphics*, 13:43–72, 1999.
66. E. Oñate and J.M. Carbonell. Updated lagrangian finite element formulation for quasi and fully incompressible fluids. *Computational Mechanics*, 54 (6), 2014.
67. P. Ryzhakov, E. Oñate, and S.R. Idelsohn. Improving mass conservation in simulation of incompressible flows. *International Journal of Numerical Methods in Engineering*, 90:1435–1451, 2012.

68. P. Ryzhakov, J. Cotela, R. Rossi, and E. Oñate. A two-step monolithic method for the efficient simulation of incompressible flows. *International Journal for Numerical Methods in Fluids*, 74 (12):919–934, 2014.
69. M. Benzi, G.H. Golub, and J. Liesen. Numerical solution of saddle point problems. *Acta Numerica*, 14 (1):1–137, 2005.
70. M. Benzi, M.A. Olshanskii, and Z. Wang. Modified augmented lagrangian preconditions for the incompressible navier-stokes equations. *International Journal for Numerical Methods in Fluids*, 66:486–508, 2011.
71. H.C. Elman, D.J. Silvester, and A.J. Wathen. Finite element and fast iterative solvers with applications in incompressible fluid dynamics. *Oxford Series in Numerical Mathematics and Scientific Computation. Oxford University Press, Oxford*, 2005.
72. M. Fortin and R. Glowinski. Augmented lagrangian: application to the numerical solution of boundary value problems. *North-Holland, Amsterdam*, 1983.
73. S. Vincent, Al Sarthou, J.P. Caltagirone, F. Sonilhac, P. Février, C. Mignot, and G. Pianet. Augmented lagrangian and penalty methods for the simulation of two-phase flows interacting with moving solids. application to hydroplanning flows interacting real tire tread patterns. *Journal of Computational Physics*, 203 (4):956–983, 2013.
74. D. Bresch, E.D. Fernández Nieto, I.R. Ionescu, and P. Vigneaux. Augmented lagrangian method and compressible visco-plastic flow. applications to shallow dense avalanches. *New Directions in Mathematical Fluid Mechanics. The Alexander V. Kazhikhov Memorial Volume. In the Series of Book on Advances in Mathematical Fluid Mechanics. A.V. Fursikov, G.P. Galdi and V.V. Pukhnachev (Eds.), Birkhauser Verlag Basel*, pages 57–89, 2010.
75. G. Vinay, A. Wachs, and J.F. Agassant. Numerical simulation of weakly compressible bingham flows: The restart of pipeline flows of waxy crude oil. *Journal of Non-Newtonian Fluids Mechanics*, 136 (2–3):93–105, 2006.

Chapter 2
Velocity-Based Formulations for Compressible Materials

In this chapter two velocity-based finite element formulations for compressible materials are presented, namely the Velocity (V) and the mixed Velocity–Pressure (VP) formulations. For both schemes the linear momentum equations are solved iteratively for the velocity increments. The linearization of the governing equations is performed without specifying any constitutive law. The aim of this chapter is to maintain as much as possible the generality of the algorithms, leaving the formulations open to different material models. It will be shown that the only requirement demanded to the constitutive laws is that the rate of stress must be linearly related with the rate of deformation.

There are several reasons that justify the presentation of the Velocity formulation. First of all, the tangent matrix of the linear momentum equations for the Velocity formulation holds also for the here proposed mixed Velocity–Pressure formulation. Furthermore, the Velocity formulation is useful for making some interesting and didactic comparisons with the mixed formulations.

After deriving the solution scheme for the Velocity formulation, the mixed Velocity–Pressure method is presented. The governing equations are the linear momentum and the linear pressure-deformation rate equations. The latter is called continuity, or mass balance, equation for its similarity with the incompressibility constraint of the Navier–Stokes problem. As for the velocity scheme, in the mixed formulation the constitutive law is not specified. In Sect. 2.3 the hypoelastic–plastic model is presented and this constitutive law is inserted in both the Velocity and the mixed Velocity–Pressure schemes. The incremental solution scheme is explained in detail for both formulations. At the end of the chapter some validation examples for hypoelastic–plastic solids in statics as in dynamics are given.

© Springer International Publishing AG 2017
A. Franci, *Unified Lagrangian Formulation for Fluid and Solid Mechanics,*
Fluid-Structure Interaction and Coupled Thermal Problems Using the PFEM,
Springer Theses, DOI 10.1007/978-3-319-45662-1_2

2.1   Velocity Formulation

In this section, the Velocity formulation for solving transient problems for a general continuum is derived. The governing equations are the linear momentum equations and they are derived in the updated Lagrangian (UL) framework. This means that the governing equations are integrated over the unknown configuration Ω (the so-called updated configuration). As a consequence, the space derivatives for the UL description are computed with respect to the spatial coordinates.

2.1.1   From the Local Form to the Spatial Semi-discretization

In this section the spatial semi-discretization of the linear momentum equations is derived.

For a general continuum, the local form of the linear momentum equations using the UL description reads

$$\rho(\mathbf{X}, t)\frac{\partial v(\mathbf{X}, t)}{\partial t} - \frac{\partial \sigma(\mathbf{X}, t)}{\partial \mathbf{x}} - b(\mathbf{X}, t) = 0 \qquad in \ \Omega \times (0, T) \qquad (2.1)$$

where ρ is the density of the material, v are the velocity vector, σ is the Cauchy stress tensor and b is the body force vector. The variables within the brackets are the independent variables: $\mathbf{X}$ are the Lagrangian or material coordinates vector, $\mathbf{x}$ the Eulerian or spatial coordinates vector and t is the time. For simplicity, in what follows the independent variables are not specified. The spatial and material coordinates are related through the motion tensor $\boldsymbol{\Phi}$ as

$$\mathbf{x} = \boldsymbol{\Phi}(\mathbf{X}, t), \qquad \mathbf{X} = \boldsymbol{\Phi}^{-1}(\mathbf{x}, t) \qquad (2.2)$$

The set of governing equations is completed by the following conditions at the Dirichlet (Γ_v) and Neumann (Γ_t) boundaries

$$v_i - v_i^p = 0 \qquad on \ \Gamma_v \qquad (2.3)$$

$$\sigma_{ij}n_j - t_i^p = 0 \qquad on \ \Gamma_t \qquad (2.4)$$

where v_i^p and t_i^p are the prescribed velocities and the prescribed tractions, respectively, and n is the unit normal vector.

In the following summation of terms for repeated indices is assumed, unless otherwise specified.

The spaces for the trial and test functions are defined, respectively, as

$$v_i \in U, \qquad U = \{v_i | v_i \in C^0, \ v_i = v_i^p \ on \ \Gamma_v\} \qquad (2.5)$$

$$w_i \in U_0, \qquad U_0 = \{w_i | w_i \in C^0,\ w_i = 0\ on\ \Gamma_v\} \tag{2.6}$$

Multiplying Eqs. (2.1) by the test functions and integrating over the updated configuration domain, the following global integral form is obtained

$$\int_\Omega w_i \left(\rho \dot{v}_i - \frac{\partial \sigma_{ij}}{\partial x_j} - b_i \right) d\Omega = 0 \tag{2.7}$$

where the symbol $(\dot{\cdot})$ represents the material time derivative.

Integrating by parts the term involving σ_{ij} in Eq. (2.7) and using the Neumann boundary conditions (2.4) yields the weak variational form of the momentum equations as

$$\int_\Omega w_i \rho \dot{v}_i d\Omega + \int_\Omega \frac{\partial w_i}{\partial x_j} \sigma_{ij} d\Omega - \int_\Omega w_i b_i d\Omega - \int_{\Gamma_t} w_i t_i^p d\Gamma = 0 \tag{2.8}$$

Equation (2.8) is the standard form of the Principle of Virtual Power [1].

The spatial discretization is introduced using the classical FEM-Galerkin procedure [2]. Hence both the trial and the test functions are interpolated in space in terms of their nodal values by means of the same shape functions N

$$v_i = \sum_{I=1}^n N_I(X) \bar{v}_{iI}\ , \quad w_i = \sum_{I=1}^n N_I(X) \bar{w}_{iI} \tag{2.9}$$

where, assuming the use of simplicial elements, $n = 3/4$ for 2D/3D problems is the number of the nodes of the element, $(\bar{\cdot})$ denotes a nodal value, the capital subscript specifies the node and the lower case subscript represents the cartesian direction. In this work, linear shape functions have been used for N_I.

Since Eq. (2.9) must hold for all the test functions in the interpolation space, introducing the spatial discretization (2.9) into Eq. (2.8), the spatial semi-discretized form of the momentum equations in the UL framework for the node I reads

$$\underbrace{\int_\Omega N_I \rho d\Omega\ \dot{v}_i}_{f_{Ii}^{dyn}} + \underbrace{\int_\Omega \frac{\partial N_I}{\partial x_j} \sigma_{ij} d\Omega}_{f_{Ii}^{int}} = \underbrace{\int_\Omega N_I b_i d\Omega + \int_{\Gamma_t} N_I t_i^p d\Gamma}_{f_{Ii}^{ext}} \tag{2.10}$$

where f^{dyn}, f^{int} and f^{ext} are the dynamic, internal and external forces, respectively, expressed in the UL framework.

For convenience, the semi-discretized form of the momentum equations in the total Lagrangian (TL) framework is also presented here. This is written as [1]

$$\underbrace{\int_{\Omega_0} N_I \rho_0 d\Omega \, \dot{v}_i}_{TL\,f_{Ii}^{dyn}} + \underbrace{\int_{\Omega_0} \frac{\partial N_I}{\partial X_j} P_{ij} d\Omega}_{TL\,f_{Ii}^{int}} = \underbrace{\int_{\Omega_0} N_I b_i d\Omega + \int_{\Gamma_0} N_I t_{0i}^p d\Gamma}_{TL\,f_{Ii}^{ext}} \qquad (2.11)$$

where P is the first Piola–Kirchhoff stress tensor, or the nominal stress tensor, and $TL\,f^{dyn}$, $TL\,f^{int}$ and $TL\,f^{ext}$ are the dynamic, internal and external forces, respectively, expressed in the TL framework. All the variables with vectors subscript $(\cdot)_0$ refer to the last known configuration. Note that Eq. (2.11) can be obtained from Eq. (2.10) by pull back transformations on all its terms [1].

For the sake of clarity in the notation, the terms referred to the TL description are denoted with the left index $TL(\cdot)$. Unless otherwise specified, the variables belong to the UL description.

2.1.2 Time Integration

In this work, the kinematic variables have been integrated in time using a second order scheme. In particular, the implicit Newmark's integration rule has been adopted. For the general case, this rule states that accelerations and displacements are computed as

$$\begin{aligned} {}^{n+1}\dot{v} &= \frac{1}{\gamma \Delta t} \left({}^{n+1}v - {}^n v \right) - \frac{1-\gamma}{\gamma} {}^n \dot{v} \\ {}^{n+1}u &= {}^n u + \Delta t \frac{\gamma - \beta}{\gamma} {}^n v + \Delta t \frac{\beta}{\gamma} {}^{n+1}v + \Delta t^2 \frac{\gamma - 2\beta}{2\gamma} {}^n \dot{v} \end{aligned} \qquad (2.12)$$

where β and γ are the so-called Newmark's parameters [1]. This time integration scheme is unconditionally stable if the following relation holds

$$\gamma \geq 2\beta \geq \frac{1}{2} \qquad (2.13)$$

In the present work, the Newmark's parameters chosen are $\beta = \frac{1}{4}$ and $\gamma = \frac{1}{2}$.
Replacing the numerical values of the constants in Eq. (2.12) yields

$$^{n+1}\dot{v} = \frac{2}{\Delta t} \left({}^{n+1}v - {}^n v \right) - {}^n \dot{v} \qquad (2.14)$$

$$^{n+1}u = {}^n u + \frac{\Delta t}{2} \left({}^{n+1}v + {}^n v \right) \qquad (2.15)$$

2.1.3 Linearization

Although the problem is set out in the UL framework, the linearization for the velocities increment of the momentum equations is performed first on the TL semi-discretized form (2.11). The UL linearized form is obtained by performing a push-forward transformation on the TL form. This is justified by the easier derivation of the tangent matrix in the TL framework. In fact, in Eq. (2.11) the only variable that depends on time is the nominal stress P, while in the UL form (2.10) the time-dependent variables are the updated domain Ω, the Cauchy stress tensor σ and the spatial derivatives $\partial N / \partial x$. For the sake of clarity, the linearization of the internal and dynamic forces will be performed separately.

Internal Component of the Tangent Matrix

From Eq. (2.11) the internal forces in the TL description are defined as

$$^{TL} f_{Ii}^{int} = \int_{\Omega_0} \frac{\partial N_I}{\partial X_j} P_{ij} d\Omega_0 \tag{2.16}$$

The constitutive relation is expressed in rate form. Hence it is more convenient to perform the linearization of the material derivative of the internal forces and then integrate for the time step increment Δt. The material time derivative of (2.16) is

$$^{TL} \dot{f}_{Ii}^{int} = \int_{\Omega_0} \frac{\partial N_I}{\partial X_j} \dot{P}_{ij} d\Omega_0 \tag{2.17}$$

The infinitesimal increment of Eq. (2.17) is

$$^{TL} \delta \dot{f}_{Ii}^{int} = \int_{\Omega_0} \frac{\partial N_I}{\partial X_j} \delta \dot{P}_{ij} d\Omega_0 \tag{2.18}$$

The first Piola–Kirchhoff stress tensor P is not typically used because it is not symmetric and its rate is a non-objective measure. For these reasons, in the TL framework it is more convenient to work with the second Piola–Kirchhoff stress tensor S and its rate. These stress rate measures are related each other through the following relation

$$\dot{P}_{ij} = \dot{S}_{ir} F_{rj}^T + S_{ir} \dot{F}_{rj}^T \tag{2.19}$$

where F is the deformation gradient tensor defined as

$$F_{ij} = \frac{\partial x_i}{\partial X_j} \tag{2.20}$$

Substituting Eq. (2.19) into (2.18) yields

$$^{TL}\delta \dot{f}_{Ii}^{int} = \underbrace{\int_{\Omega_0} \frac{\partial N_I}{\partial X_j} F_{ir} \delta \dot{S}_{jr} d\Omega_0}_{^{TL}\delta \dot{f}_{Ii}^{m}} + \underbrace{\int_{\Omega_0} \frac{\partial N_I}{\partial X_j} S_{jr} \delta \dot{F}_{ir}^{T} d\Omega_0}_{^{TL}\delta \dot{f}_{Ii}^{g}} \qquad (2.21)$$

Equation (2.21) shows that the increment of the material time derivative of the internal forces can be split into material and geometric parts, $^{TL}\delta \dot{\boldsymbol{f}}^{m}$ and $^{TL}\delta \dot{\boldsymbol{f}}^{g}$, respectively. The former accounts for the material response through the rate of the second Piola–Kirchhoff stress tensor and the second term is the initial stress term that contains the information of the updated stress field. Note that, up to now, no constitutive relationships have been introduced and the above derivation holds for a general continuum.

Material Tangent Matrix

From Eq. (2.21), the material part of the material time derivative of the internal forces reads

$$^{TL}\delta \dot{f}_{Ii}^{m} = \int_{\Omega_0} \frac{\partial N_I}{\partial X_j} F_{ir} \delta \dot{S}_{jr} d\Omega_0 \qquad (2.22)$$

For the derivation of the material tangent matrix, the constitutive relation between the stress and the strain measures is required. In order to maintain the formulation as general as possible, the stress rate measure is related to the deformation rate through a tangent constitutive tensor, as

$$\dot{S}_{ij} = C_{ijkl} \dot{E}_{kl} \qquad (2.23)$$

where $\boldsymbol{C}$ is a fourth-order tensor and $\boldsymbol{E}$ is the Green–Lagrange strain tensor. Equation (2.23) can also be expressed in Voigt notation as

$$\{\dot{S}\} = [C]\{\dot{E}\} \qquad (2.24)$$

where $\{(\cdot)\}$ denotes a vector with components $[(\cdot)_{11}, (\cdot)_{22}, (\cdot)_{33}, (\cdot)_{12}, (\cdot)_{13}, (\cdot)_{23}]$. As it will be shown in the following sections, Eq. (2.23) can represent both a Kirchhoff solid material and a Newtonian fluid. If different constitutive laws are used, Eq. (2.23) should be modified accordingly in order to derive the material part of the tangent matrix.

Using the Voigt notation, the Green–Lagrange strain tensor can be expressed in terms of the nodal velocities as

$$\{\dot{E}\} = \boldsymbol{B}_0 \bar{\boldsymbol{v}} \qquad (2.25)$$

where for a plane strain problem

$$\boldsymbol{B}_0 = \begin{bmatrix} \dfrac{\partial N_I}{\partial X}\dfrac{\partial x}{\partial X} & \dfrac{\partial N_I}{\partial X}\dfrac{\partial y}{\partial X} \\[2ex] \dfrac{\partial N_I}{\partial Y}\dfrac{\partial x}{\partial Y} & \dfrac{\partial N_I}{\partial Y}\dfrac{\partial y}{\partial Y} \\[2ex] \dfrac{\partial N_I}{\partial X}\dfrac{\partial x}{\partial Y} + \dfrac{\partial N_I}{\partial Y}\dfrac{\partial x}{\partial X} & \dfrac{\partial N_I}{\partial X}\dfrac{\partial y}{\partial Y} + \dfrac{\partial N_I}{\partial Y}\dfrac{\partial y}{\partial X} \end{bmatrix} \tag{2.26}$$

Substituting Eq. (2.25) in (2.24)

$$\{\dot{\boldsymbol{S}}\} = [\boldsymbol{C}]\,\boldsymbol{B}_0\boldsymbol{v} \tag{2.27}$$

Substituting Eq. (2.27) into (2.22) yields

$$^{TL}\delta \dot{\boldsymbol{f}}^m = \int_{\Omega_0} \boldsymbol{B}_0^T [\boldsymbol{C}]\,\boldsymbol{B}_0 d\Omega_0 \delta \bar{\boldsymbol{v}} \tag{2.28}$$

In order to obtain the increment of the internal forces, the material time derivative of the internal forces increment is integrated over a time step increment Δt as

$$^{TL}\delta \boldsymbol{f}^m = {}^{TL}\delta \dot{\boldsymbol{f}}^m \Delta t \tag{2.29}$$

From Eqs. (2.29) and (2.28)

$$^{TL}\delta \boldsymbol{f}^m = \int_{\Omega_0} \boldsymbol{B}_0^T \Delta t\, [\boldsymbol{C}]\,\boldsymbol{B}_0 d\Omega_0 \delta \bar{\boldsymbol{v}} \tag{2.30}$$

From Eq. (2.30), the material tangent matrix in the TL description can be computed as

$$^{TL}\boldsymbol{K}^m = \int_{\Omega_0} \boldsymbol{B}_0^T \Delta t\, [\boldsymbol{C}]\,\boldsymbol{B}_0 d\Omega_0 \tag{2.31}$$

The material tangent matrix for the UL framework is obtained by applying a push-forward transformation on each term of Eq. (2.31) and integrating over the updated domain Ω. The following relations hold

$$d\Omega_0 = \frac{d\Omega}{J} \tag{2.32}$$

$$\frac{\partial N_I}{\partial X_j} = \frac{\partial N_I}{\partial x_k} F_{kj} \tag{2.33}$$

$$C_{ijkl} = F_{mi}^{-1} F_{nj}^{-1} F_{ok}^{-1} F_{pl}^{-1} c_{mnop}^\tau = F_{mi}^{-1} F_{nj}^{-1} F_{ok}^{-1} F_{pl}^{-1} c_{mnop}^\sigma J \tag{2.34}$$

where c^τ is the tangent moduli corresponding to the material time derivative of the Kirchhoff stress tensor $\tau^\triangledown$ and c^σ is the tangent moduli for the rate of the Cauchy stress $\sigma^\triangledown$. The rate of the Cauchy stress tensor is related to the rate of deformation d through the fourth-order tensor c^σ by the following expression

$$\sigma^\triangledown = c^\sigma : d \tag{2.35}$$

Substituting Eqs. (2.32)–(2.34) into (2.31) and using the minor symmetries, the material tangent matrix for the UL reads

$$K_{IJ}^m = \int_{\Omega^e} B_I^T \Delta t \left[c^\sigma \right] B_J d\Omega \tag{2.36}$$

For the node I of a 3D element, matrix B is

$$B_I = \begin{bmatrix} \dfrac{\partial N_I}{\partial x} & 0 & 0 \\[2mm] 0 & \dfrac{\partial N_I}{\partial y} & 0 \\[2mm] 0 & 0 & \dfrac{\partial N_I}{\partial z} \\[2mm] \dfrac{\partial N_I}{\partial y} & \dfrac{\partial N_I}{\partial x} & 0 \\[2mm] \dfrac{\partial N_I}{\partial z} & 0 & \dfrac{\partial N_I}{\partial x} \\[2mm] 0 & \dfrac{\partial N_I}{\partial z} & \dfrac{\partial N_I}{\partial y} \end{bmatrix} \tag{2.37}$$

Geometric Tangent Matrix

The geometric tangent matrix for the UL framework is derived next using the same procedure adopted for the material components. Hence, first the linearization is performed using the TL form and then the UL tangent matrix is obtained by performing the required transformation over the TL terms.

From Eq. (2.21)

$$^{TL}\delta \dot{f}_{Ii}^g = \int_{\Omega_0} \frac{\partial N_I}{\partial X_j} S_{jr} \delta \dot{F}_{ir}^T d\Omega_0 \tag{2.38}$$

where the rate of the deformation gradient is defined as

$$\dot{F}_{ij} = \frac{\partial N_J}{\partial X_j} \bar{v}_{Ji} \tag{2.39}$$

Substituting Eq. (2.39) into (2.38), the geometric components of the internal power in the TL description can be written as

$$^{TL}\delta \dot{f}_{Ii}^{g} = \int_{\Omega_0} \frac{\partial N_I}{\partial X_j} S_{jr} \frac{\partial N_J}{\partial X_r} d\Omega_0 \, \delta \bar{v}_{Ji} \tag{2.40}$$

Integrating Eq. (2.40) on time for a time step increment Δt yields

$$^{TL}\delta f_{Ii}^{g} = \int_{\Omega_0} \frac{\partial N_I}{\partial X_j} \Delta t S_{jr} \frac{\partial N_J}{\partial X_r} d\Omega \, \delta_{ik}\delta \bar{v}_{Jk} \tag{2.41}$$

In order to recover the UL form, the Piola identity has to be recalled, i.e.

$$S = F^{-1} \sigma F^{-T} J \tag{2.42}$$

The geometric tangent matrix in the UL framework is obtained by substituting Eqs. (2.32), (2.33) and (2.42) into (2.41) and using the symmetries. This leads to

$$K_{IJik}^{g} = \int_{\Omega} \frac{\partial N_I}{\partial x_j} \Delta t \sigma_{jr} \frac{\partial N_J}{\partial x_r} d\Omega \delta_{ik} \tag{2.43}$$

or also

$$\boldsymbol{K}_{IJ}^{g} = \boldsymbol{I} \int_{\Omega} \boldsymbol{\beta}_{I}^{T} \Delta t \sigma \boldsymbol{\beta}_{J} d\Omega \tag{2.44}$$

where $\boldsymbol{I}$ is the second order identity tensor and matrix β for 3D problems is

$$\boldsymbol{\beta}_I = \left[\frac{\partial N_I}{\partial x} \; \frac{\partial N_I}{\partial y} \; \frac{\partial N_I}{\partial z} \right]^T \tag{2.45}$$

Dynamic Component of the Tangent Matrix

The dynamic component of the tangent matrix in the UL description can be derived directly from the UL dynamic term f_{Ii}^{dyn} of Eq. (2.10). This reads

$$f_{Ii}^{dyn} = \int_{\Omega} N_I \rho d\Omega \, \dot{v}_i \tag{2.46}$$

Equation (2.46) has to be discretized in time with the purpose of replacing the accelerations by the velocities using the time integration scheme described in Eq. (2.14).

Introducing Eq. (2.14) into (2.46) and differentiating for the increment of velocities, the dynamic components of the tangent matrix are obtained as

$$K^{\rho}_{IJij} = \delta_{ij} \int_{\Omega} N_I \frac{2\rho}{\Delta t} N_J d\Omega \qquad (2.47)$$

Or also

$$K^{\rho}_{IJ} = I \int_{\Omega} N_I \frac{2\rho}{\Delta t} N_J d\Omega \qquad (2.48)$$

2.1.4 Incremental Solution Scheme

The problem is solved through an implicit iterative scheme. At each iteration i the velocity increments are computed as

$$K^i \Delta \bar{v} = R^i \left({}^{n+1}\bar{v}^i, {}^{n+1}\sigma^i \right) \qquad (2.49)$$

where K is the tangent matrix computed as the sum of the internal, the geometric and the dynamic components, given by Eqs. (2.36), (2.44) and (2.48) respectively, as

$$K = K^m + K^g + K^\rho \qquad (2.50)$$

R is the residual and it is computed from Eq. (2.10) as

$$^{n+1}R_{Ii} = \int_{\Omega} N_I \rho N_J d\Omega \, {}^{n+1}\bar{v}_{Ji} + \int_{\Omega} \frac{\partial N_I}{\partial x_j} {}^{n+1}\sigma_{ij} d\Omega - \int_{\Omega} N_I {}^{n+1}b_i d\Omega +$$
$$- \int_{\Gamma_t} N_I {}^{n+1}t_i^p d\Gamma \quad (2.51)$$

In Eq. (2.51) $^{n+1}R_{Ii}$ is the residual of the momentum equations referred to node I and the cartesian direction i. Note that the Cauchy stress tensor still appears in its 'continuum' form because up to now it has not been written as a function of the nodal unknowns. This is done for keeping the generality of the formulation. Only after the introduction of the constitutive laws, it will be possible to compute the Cauchy stress tensor as a function of the nodal unknowns.

In Box 1 the iterative solution incremental scheme of the velocity formulation for a generic time interval $[{}^nt, {}^{n+1}t]$ of duration Δt is described.

For each iteration i within a time interval:

1. Compute the nodal velocity increments $\Delta \bar{v}$:

$$K^i \Delta \bar{v} = R^i \left({}^{n+1}\bar{v}^i, \sigma({}^{n+1}\bar{v}^i) \right)$$

where: $K^i = K^m \left({}^{n+1}\bar{x}^i, c^{\sigma,i} \right) + K^g \left({}^{n+1}\bar{x}^i, \sigma(\bar{v}^i) \right) + K^p \left({}^{n+1}\bar{x}^i \right)$

2. Update the nodal velocities: ${}^{n+1}\bar{v}^{i+1} = {}^{n+1}\bar{v}^i + \Delta \bar{v}$

3. Update the nodal coordinates: ${}^{n+1}\bar{x}^{i+1} = {}^{n+1}\bar{x}^i + \bar{u}(\Delta \bar{v})$

4. Compute the updated stress tensors: $\sigma({}^{n+1}\bar{v}^{i+1})$ and $\sigma^{\nabla}({}^{i+1}\bar{v}^{i+1})$

5. Check convergence: $\| {}^{n+1}R^{i+1}({}^{n+1}\bar{v}^{i+1}) \| < tolerance$

If condition 5 is not fulfilled, return to 1 with $i \leftarrow i + 1$.

Box 1. Iterative incremental solution scheme for the velocity formulation

2.2 Mixed Velocity–Pressure Formulation

In this work, the mixed Velocity–Pressure formulation is derived as an extension of the Velocity formulation presented in the previous section. The governing equations are the linear momentum equations and the linear relation between the time variation of pressure and the volumetric strain rate.

The problem is solved using a two-step Gauss–Seidel partitioned iterative scheme. First the momentum equations are solved in terms of velocity increments and including the (known) pressures at the previous iteration in the residual expression. Then the continuity equation is solved for the pressure using the updated velocities computed from the momentum equations. It will be shown that using this not intrusive scheme, it is possible to take advantage of most of the velocity formulation derived in the previous section. In particular, the incremental velocity scheme for the momentum equations (Box 1) and the structure of the tangent matrix (2.50) hold also for the mixed Velocity–Pressure formulation. The same linear interpolation has been used for the velocity and the pressure fields. It is well known that, for incompressible (or quasi-incompressible) problems, this combination does not fulfill the *inf–sup* condition [3] and a stabilization method is required. However, as mentioned in the section devoted to the velocity formulation, the aim of this part is to keep the formulation as general as possible without referring to a specific material. Hence only in the next chapter, when the mixed Velocity–Pressure formulation is used for solving quasi-incompressible problems, the required stabilization will be introduced in the scheme.

2.2.1   Quasi-incompressible Form of the Continuity Equation

Mixed formulations are often used for dealing with incompressible materials. In these problems it is useful to write the stress and the strain measures as the sum of deviatoric and hydrostatic, or volumetric, parts. Hence the Cauchy stress tensor is decomposed as

$$\sigma_{ij} = \sigma'_{ij} + \sigma^h \delta_{ij} \tag{2.52}$$

with

$$\sigma^h = \frac{\sigma_{kk}}{3} \tag{2.53}$$

where $\boldsymbol{\sigma}'$ and $\sigma^h \boldsymbol{I}$ are the deviatoric and the hydrostatic parts of the Cauchy stress tensor, respectively. The pressure p is defined positive in the tensile state and equal to the hydrostatic parts of the Cauchy stress tensor σ^h. Hence

$$p := \sigma^h \tag{2.54}$$

The Cauchy stress tensor can be computed as

$$\sigma_{ij} = \sigma'_{ij} + p\delta_{ij} \tag{2.55}$$

The same decomposition is done for the spatial strain rate tensor $\boldsymbol{d}$. So

$$d_{ij} = d'_{ij} + d^h \delta_{ij} \tag{2.56}$$

with

$$d^h = \frac{d_{kk}}{3} \tag{2.57}$$

where $\boldsymbol{d}'$ and $d^h \boldsymbol{I}$ are the deviatoric and the hydrostatic parts of the strain rate tensor, respectively. The strain rate tensor is computed from the velocities as

$$d_{ij} = \frac{1}{2}\left(\frac{\partial v_i}{\partial x_j} + \frac{\partial v_j}{\partial x_i} \right) \tag{2.58}$$

The volumetric strain rate is defined from Eqs. (2.57) and (2.58) as

$$d^v = d_{kk} = \frac{\partial v_k}{\partial x_k} \tag{2.59}$$

The closure equation for the mixed Velocity–Pressure formulation is the linear relation between the change in time of the pressure and the volumetric strain rate. This reads as

$$\frac{1}{\kappa}\frac{\partial p}{\partial t} = d^v \tag{2.60}$$

where κ is a parameter that depends on the constitutive equation. Typically κ is the bulk modulus of the material.

In conclusion, the local form of the whole problem for the mixed Velocity–Pressure formulation is formed by the linear momentum equations (Eq. (2.1)) and the pressure-strain rate relation given by Eq. (2.60). The linear momentum equations have been already discretized and linearized for the increments of velocities in the previous section devoted to the Velocity formulation. That form holds also for the mixed formulation. So, in the following, only the discretization of Eq. (2.60) is given. This equation is a restriction on the generality of the constitutive laws that can be analyzed with the mixed Velocity–Pressure formulation. It will be shown that the constitutive models for hypoelastic solids and quasi-incompressible Newtonian fluids fulfill this relation. In fluid dynamics, Eq. (2.60) represents the *continuity*, or *mass balance*, equation for quasi-incompressible fluids. In fact, Eq. (2.60) with $\kappa = \infty$ is the canonical form of the continuity equation of the Navier–Stokes problem. For this reason, from here on Eq. (2.60) will be called 'continuity equation'.

Multiplying Eq. (2.60) by arbitrary test functions q (with dimensions of pressure), integrating over the analysis domain Ω and bringing all the terms at the left hand side gives

$$\int_{\Omega}\frac{q}{\kappa}\frac{\partial p}{\partial t}d\Omega - \int_{\Omega}qd^v d\Omega = 0 \tag{2.61}$$

Both the trial and the test functions for the pressure are interpolated in space using the same shape functions N.

$$p = \sum_{I=1}^{n}N_I\bar{p}_I \quad , \quad q = \sum_{I=1}^{n}N_I\bar{q}_I \tag{2.62}$$

where $n = 3/4$ for 2D/3D problems is the number of the nodes of the simplex. In this work, linear shape functions have been used for N_I, as for the velocity.

Combining Eq. (2.62) with (2.61) and solving for all the admissible test functions q, yields

$$\int_{\Omega}N_I\frac{1}{\kappa}N_Jd\Omega\dot{\bar{p}}_J - \int_{\Omega}N_I\frac{\partial N_J}{\partial x_i}d\Omega\bar{v}_{iJ} = 0 \tag{2.63}$$

Regarding the time integration a first order scheme has been adopted for the pressure. Hence, for a time interval $[^nt, {}^{n+1}t]$ of duration Δt the first and the second variations in time of the pressure are computed as

$$^{n+1}\dot{p} = \frac{^{n+1}p - {}^{n}p}{\Delta t}$$

(2.64)

$$^{n+1}\ddot{p} = \frac{^{n+1}p - {}^{n}p}{\Delta t^2} - \frac{^{n}\dot{p}}{\Delta t}$$

(2.65)

Introducing Eq. (2.64) in Eq. (2.63), the discretized form of the continuity equation is

$$-Q^{T\,n+1}\bar{v} + \frac{1}{\Delta t}M_1{}^{n+1}\bar{p} = {}^{n}g$$

(2.66)

where

$$M_{1_{IJ}} = \int_{\Omega^e} N_I \frac{1}{\kappa} N_J d\Omega$$

(2.67)

$$Q_{IJ} = \int_{\Omega^e} B_I^T m N_J d\Omega$$

(2.68)

$$^{n}g_I = \int_{\Omega^n} N_I \frac{1}{\kappa \Delta t} N_J d\Omega \, {}^{n}\bar{p}_J$$

(2.69)

where B has been defined in Eq. (2.37) and $m = [1, 1, 1, 0, 0, 0]^T$.

2.2.2   Solution Method

In the mixed Velocity–Pressure formulation the problem is solved through a partitioned iterative scheme. Specifically, the linear momentum equations are solved for the velocity increments as in the Velocity formulation. On the other hand, the continuity equation is solved for the pressure in the updated configuration using the velocity field computed with the linear momentum equations. In order to guarantee the coupling between the continuity equation and the linear momentum equations (or equally between the pressure and the velocities) the pressure must appear in the right hand side of the linear momentum equations. For this purpose the Cauchy stress tensor must be computed as the sum of its deviatoric part and the pressure, as Eq. (2.55). Otherwise, the Velocity–Pressure formulation would be uncoupled and totally equivalent to the Velocity formulation.

In conclusion, for a general time interval $[{}^{n}t, {}^{n+1}t]$ of duration Δt the following linear systems are solved for each iteration i

$$K^i \Delta \bar{v} = R^i ({}^{n+1}\bar{v}^i, {}^{n+1}\sigma'^i, {}^{n+1}p^i)$$

(2.70)

$$\frac{1}{\Delta t}M_1{}^{n+1}\bar{p}^{i+1} = Q^{T\,n+1}\bar{v}^{i+1} + {}^{n}g^{n}\bar{p}^i$$

(2.71)

where K is the same tangent matrix as for the Velocity formulation (Eq. (2.50)) and the residual R is computed using the pressure of the previous iteration and the deviatoric part of the Cauchy stress as

$$^{n+1}R_{Ii} = \int_{\Omega} N_I \rho N_J d\Omega \, ^{n+1}\bar{v}_{Ji} + \int_{\Omega} \frac{\partial N_I}{\partial x_j} \, ^{n+1}\sigma'_{ij} d\Omega +$$

$$+ \int_{\Omega} \frac{\partial N_I}{\partial x_j} \delta_{ij} N_J d\Omega \, ^{n+1}\bar{p}_J - \int_{\Omega} N_I \, ^{n+1}b_i d\Omega - \int_{\Gamma_t} N_I \, ^{n+1}t_i^p d\Gamma \qquad (2.72)$$

In Box 2, the iterative incremental solution scheme for a generic continuum via the mixed Velocity–Pressure formulation is shown for a time interval $[^n t, \, ^{n+1}t]$.

For each iteration i within a time interval:

1. Compute the nodal velocity increments $\Delta \bar{v}$:

$$K^i \Delta \bar{v} = R^i (^{n+1}\bar{v}^i, \, ^{n+1}\bar{p}^i)$$

where: $K^i = K^m (^{n+1}\bar{x}^i, c^{\sigma,i}) + K^g (^{n+1}\bar{x}^i, \sigma^i(\bar{v}^i, \bar{p}^i)) + K^p(^{n+1}\bar{x}^i)$

2. Update the nodal velocities: $\quad ^{n+1}\bar{v}^{i+1} = {}^{n+1}\bar{v}^i + \Delta \bar{v}$

3. Update the nodal coordinates: $\quad ^{n+1}\bar{x}^{i+1} = {}^{n+1}\bar{x}^i + \bar{u}(\Delta \bar{v})$

4. Compute the nodal pressures $^{n+1}\bar{p}^{i+1}$:

$$\frac{1}{\Delta t} M_1 {}^{n+1}\bar{p}^{i+1} = Q^T \, ^{n+1}\bar{v}^{i+1} + g^n(^n\bar{p}^i)$$

5. Compute the updated stress measures:

$$^{n+1}\sigma^{i+1} = {}^{n+1}\sigma'(\bar{v}^{i+1}) - {}^{n+1}p^{i+1}I$$

6. Check convergence: $\quad \| {}^{n+1}R^{i+1}(^{n+1}\bar{v}^{i+1}, \, ^{n+1}\bar{p}^{i+1}) \| < tolerance$

If condition 6 is not fulfilled, return to 1 with $i \leftarrow i + 1$.

Box 2. Iterative solution scheme for a generic continuum using the mixed Velocity–Pressure formulation

2.3 Hypoelasticity

Using the definition of Truesdell [4], a hypoelastic body is a material which may soften or harden in strain but in general has neither preferred state nor preferred stress. The hypoelastic laws were created with the purpose of transferring the linear theory of elasticity from the small to the finite strains regime [4]. In [5] a deep dissertation about the differences between elasticity and hypoelasticity is given.

A hypoelastic body is defined by the constitutive equation [6]

$$rate\ of\ stress = f(rate\ of\ deformation) \tag{2.73}$$

In the rate theory it is crucial to guarantee the *objectivity* and the *frame-invariance*, or *frame-indifference*, of the scheme. A material is frame invariant if its properties do not depend on the change of observer. An objective constitutive equation is defined to be invariant for all changes of the observer [7]. For guaranteeing the frame indifference, the constitutive law has to be isotropic [8]. This represents a constraint for hypoelastic models. This limitation is even more severe if also plasticity is included in the model. In fact, for hypoelastic–plastic materials, also the yield condition is required to be isotropic [9].

The stress rate cannot be computed simply as a material derivative because it leads to a non-objective measure of stress [1]. In particular, rigid rotations may originate a wrong state of stress if the stress rate is computed as the material time derivative of the Cauchy stress [1]. However, many objective measures of rate of stress are available. The most common ones are the Truesdell's and Jaumann's Cauchy stress rate measures. From here on, an objective rate measure will be denoted by the upper inverse triangle index $(\cdot)^\nabla$.

Most of hypoelastic laws relate linearly the stress rate to the rate of deformation. Hence, Eq. (2.73) is now rewritten in the following form

$$\sigma^\nabla = c : d \tag{2.74}$$

where σ^∇ is the Cauchy stress rate tensor, c is the tangent moduli tensor and d is the deformation rate tensor.

From Eq. (2.74) it can be deduced that this class of hypoelastic materials has a rate-independent and incrementally linear and reversible behavior. So, as for the elastic materials, in the small deformation regime, the strains and the stresses are totally recovered upon the unloading process. Nevertheless, for large deformations the hypoelastic laws do not guarantee that the work done in a closed deformation path is zero [10]. However this error can be considered negligible if the elastic deformations are small with respect to the total deformations [1]. For this reason the hypoelastic laws are often used for describing the elastic part of elastic-plastic materials: the plastic deformations in fact represent usually the largest part of the overall deformations.

The Jaumann measure for the rate of the Kirchhoff stress tensor $\tau^{\nabla J}$ is

$$\tau^{\nabla J} = c^{\tau J} : d \qquad (2.75)$$

where the tangent moduli fourth-order tensor $c^{\tau J}$ for the Jaumann measure of the Kirchhoff stress rate is

$$c_{ijkl}^{\tau J} = \lambda \delta_{ij}\delta_{kl} + \mu \left(\delta_{ik}\delta_{jl} + \delta_{il}\delta_{kj}\right) \quad , \quad c^{\tau J} = \lambda I \otimes I + 2\mu \mathbf{I} \qquad (2.76)$$

where λ and μ are the Lamé constants and they are computed from the Young modulus E and the Poisson ratio ν as

$$\mu = \frac{E}{2(1+\nu)} \qquad (2.77)$$

$$\lambda = \frac{\nu E}{(1+\nu)(1-2\nu)} \qquad (2.78)$$

and $\mathbf{I}$ is the fourth-order symmetric identity tensor defined as

$$\mathbf{I}_{ijkl} = \frac{1}{2}\left(\delta_{ik}\delta_{jl} + \delta_{il}\delta_{kj}\right) \qquad (2.79)$$

Separating the volumetric from the deviatoric part, yields

$$c_{ijkl}^{\tau J} = \kappa \delta_{ij}\delta_{kl} + \mu \left(\delta_{ik}\delta_{jl} + \delta_{il}\delta_{kj} - \frac{2}{3}\delta_{ij}\delta_{kl}\right) \quad , \quad c^{\tau J} = \kappa I \otimes I + 2\mu \mathbf{I}' \qquad (2.80)$$

where κ is the bulk modulus and it is computed from the Lamé parameters as

$$\kappa = \lambda + \frac{2}{3}\mu \qquad (2.81)$$

and $\mathbf{I}'$ is the fourth-order tensor computed as

$$\mathbf{I}' = \mathbf{I} - \frac{1}{3}I \otimes I \qquad (2.82)$$

The Jaumann stress rate measure is defined as

$$\sigma^{\nabla J} = c^{\sigma J} : d \qquad (2.83)$$

where $c^{\sigma J}$ is the Jaumann's tangent moduli tensor.

For a anisotropic material the tangent moduli for the Jaumann rate depends on the stress state and it is related to $c^{\tau J}$ as follows

$$c_{ijkl}^{\sigma J} = \frac{c_{ijkl}^{\tau J}}{J} - \sigma_{il}\delta_{kj} \quad , \quad \boldsymbol{c}^{\sigma J} = \frac{\boldsymbol{c}^{\tau J}}{J} - \boldsymbol{\sigma} \otimes \boldsymbol{I} \tag{2.84}$$

Instead, for isotropic materials, the Jaumann's tangent moduli tensors for the Cauchy stress rate and for the Kirchhoff stress rate are identical [1]. So $\boldsymbol{c}^{\sigma J}$ can be computed as

$$c_{ijkl}^{\sigma J} = \lambda\delta_{ij}\delta_{kl} + \mu\left(\delta_{ik}\delta_{jl} + \delta_{il}\delta_{kj}\right) \quad , \quad \boldsymbol{c}^{\sigma J} = \lambda\boldsymbol{I} \otimes \boldsymbol{I} + 2\mu\mathbf{I} \tag{2.85}$$

or equally

$$c_{ijkl}^{\sigma J} = \kappa\delta_{ij}\delta_{kl} + \mu\left(\delta_{ik}\delta_{jl} + \delta_{il}\delta_{kj} - \frac{2}{3}\delta_{ij}\delta_{kl}\right) \quad , \quad \boldsymbol{c}^{\sigma J} = \kappa\boldsymbol{I} \otimes \boldsymbol{I} + 2\mu\mathbf{I}' \tag{2.86}$$

For a 3D problem tensor $\boldsymbol{c}^{\sigma J}$ is $\boldsymbol{c}^{\sigma J} = \begin{bmatrix} \lambda+2\mu & \lambda & \lambda & 0 & 0 & 0 \\ \lambda & \lambda+2\mu & \lambda & 0 & 0 & 0 \\ \lambda & \lambda & \lambda+2\mu & 0 & 0 & 0 \\ 0 & 0 & 0 & \mu & 0 & 0 \\ 0 & 0 & 0 & 0 & \mu & 0 \\ 0 & 0 & 0 & 0 & 0 & \mu \end{bmatrix}$

or, equally, $\boldsymbol{c}^{\sigma J} = \begin{bmatrix} \kappa+\frac{4}{3}\mu & \kappa-\frac{2}{3}\mu & \kappa-\frac{2}{3}\mu & 0 & 0 & 0 \\ \kappa-\frac{2}{3}\mu & \kappa+\frac{4}{3}\mu & \kappa-\frac{2}{3}\mu & 0 & 0 & 0 \\ \kappa-\frac{2}{3}\mu & \kappa-\frac{2}{3}\mu & \kappa+\frac{4}{3}\mu & 0 & 0 & 0 \\ 0 & 0 & 0 & \mu & 0 & 0 \\ 0 & 0 & 0 & 0 & \mu & 0 \\ 0 & 0 & 0 & 0 & 0 & \mu \end{bmatrix}$

A material is said to be isotropic if its behavior is uniform in all directions, so it has no preferred orientations or directions. Many materials, such as metals and ceramics, can be modeled as isotropic for small strains [1]. From the computational point of view, an isotropic constitutive law is much easier to manage than an anisotropic one and it has a lower computational cost. Isotropic laws are preferred, for example, for their symmetry properties. In fact the anisotropic tangent moduli (Eq. (2.84)) is not symmetric while, the isotropic one (Eq. (2.86)) has both minor and major symmetries, in fact

$$minor\ symmetry \quad \leftrightarrow \quad c_{ijkl}^{\sigma J} = c_{jikl}^{\sigma J} = c_{ijlk}^{\sigma J} \tag{2.87}$$

$$major\ symmetry \quad \leftrightarrow \quad c_{ijkl}^{\sigma J} = c_{klij}^{\sigma J} \tag{2.88}$$

For all these reasons, in this work the isotropic law has been used for the hypoelastic model.

The tangent moduli $c^{\sigma J}$ will be introduced into the material part K^m of the global tangent matrix Eq. (2.36) for the Velocity and the mixed Velocity–Pressure formulations indifferently.

As it has already pointed out, the Cauchy stress rate does not coincide with the material derivative of the Cauchy stress tensor. In fact, the following relation holds between both measures

$$\dot{\sigma} = \sigma^{\nabla J} + \Omega \tag{2.89}$$

where Ω is a tensor that accounts for the rotations and it is defined as

$$\Omega = W \cdot \sigma + \sigma \cdot W^T \tag{2.90}$$

where W is the spin tensor defined as

$$W_{ij} = \frac{1}{2} \left(L_{ij} - L_{ji} \right) = \frac{1}{2} \left(\frac{\partial v_i}{\partial x_j} - \frac{\partial v_j}{\partial x_i} \right) \tag{2.91}$$

In this work the tensor Ω is computed at the end of each time step. Discretizing in time Eq. (2.89) for the time step interval $[{}^n t, {}^{n+1} t]$ and expanding the Cauchy stress rate, yields

$$\frac{{}^{n+1}\sigma - {}^n\sigma}{\Delta t} = c^{\sigma J} : {}^{n+1}d + {}^n\Omega \tag{2.92}$$

Ω can be viewed as a correction of the Cauchy stress tensor. So it can be related to the Cauchy stress tensor of the previous time step as follows

$$ {}^n\hat{\sigma} = {}^n\sigma + {}^n\Omega\Delta t \tag{2.93}$$

Replacing Eq. (2.93) in (2.92), yields

$$\frac{{}^{n+1}\sigma - {}^n\hat{\sigma}}{\Delta t} = c^{\sigma J} : {}^{n+1}d \tag{2.94}$$

Substituting in Eq. (2.94) the relation for $c^{\sigma J}$ using Eq. (2.86), yields

$$\frac{{}^{n+1}\sigma - {}^n\hat{\sigma}}{\Delta t} = \left(\kappa I \otimes I + 2\mu I' \right) : {}^{n+1}d \tag{2.95}$$

Hence,

$$\frac{{}^{n+1}\sigma - {}^n\hat{\sigma}}{\Delta t} = \underbrace{\kappa \left(I \otimes I \right) : {}^{n+1}d}_{{}^{n+1}\dot{p}} + \underbrace{2\mu I' : {}^{n+1}d}_{{}^{n+1}\dot{\sigma'}} \tag{2.96}$$

The first and the second terms of the right hand side of Eq. (2.96) represent the increment in the time step of the volumetric and deviatoric parts of the Cauchy stress tensor. From Eq. (2.96) it can be deduced that for isotropic hypoelastic solids described using the Jaumann measure, the following relation holds

$$\dot{p} = \kappa d^v \tag{2.97}$$

Equation (2.97) will be used as the closure equation of the mixed Velocity–Pressure formulation for hypoelastic solids. Note that Eq. (2.97) has the same structure as Eq. (2.60) analyzed in the previous section. From Eq. (2.97) using linear shape functions N and integrating on time the pressure with a first order scheme, the same matrix form of Eq. (2.66) obtained for a general material is obtained.

In conclusion the updated stresses can be computed using the velocities only or both the pressure and the velocities, as follows

$$^{n+1}\boldsymbol{\sigma} = {}^n\hat{\boldsymbol{\sigma}} + \Delta t \left(\kappa \boldsymbol{I} \otimes \boldsymbol{I} + 2\mu \boldsymbol{I}'\right) : {}^{n+1}\boldsymbol{d} \tag{2.98}$$

$$^{n+1}\boldsymbol{\sigma} = {}^n\hat{\boldsymbol{\sigma}} + {}^{n+1}\Delta p \boldsymbol{I} + 2\Delta t \mu \boldsymbol{I}' : {}^{n+1}\boldsymbol{d} \tag{2.99}$$

Equations (2.98) and (2.99) will be used for computing the Cauchy stress tensor in the Velocity and mixed Velocity–Pressure formulations, respectively.

2.3.1 Velocity Formulation for Hypoelastic Solids

The solution scheme for hypoelastic solids is the one derived in Sect. 2.1.4 for a general continuum. The only modifications required are the definition of the tangent moduli c^σ in matrix $\boldsymbol{K}^m$ (Eq. (2.36)) and the computation of the Cauchy stress tensor from the nodal velocities according to the hypoelastic model. This tensor appears in the geometric part of the tangent matrix $\boldsymbol{K}^g$ (Eq. (2.44)) and into the residual $\boldsymbol{R}$ (Eq. (2.51)). The tangent moduli tensor is taken from the Jaumann isotropic description and it is the tensor $c^{\sigma J}$ of Eq. (2.86). Concerning the Cauchy stresses, these are computed with Eq. (2.98). The finite element implemented with this hypoelastic velocity formulation is named V-element.

The iterative solution incremental solution scheme for hypoelastic solids using the velocity formulation for a generic time interval $[^n t, {}^{n+1} t]$ is given in Box 3.

For each iteration i within a time interval:

1. Compute the nodal velocity increments $\boldsymbol{\Delta\bar{v}}$:

$$\boldsymbol{K^i \Delta\bar{v}} = \boldsymbol{R^i}({}^{n+1}\boldsymbol{\bar{v}^i}, \boldsymbol{\sigma}({}^{n+1}\boldsymbol{\bar{v}^i}))$$

where: $\boldsymbol{K^i} = \boldsymbol{K^m}({}^{n+1}\boldsymbol{\bar{x}^i}, \boldsymbol{c^{\sigma J}}) + \boldsymbol{K^g}\left({}^{n+1}\boldsymbol{\bar{x}^i}, \boldsymbol{\sigma}(\boldsymbol{\bar{v}^i})\right) + \boldsymbol{K^\rho}({}^{n+1}\boldsymbol{\bar{x}^i})$

2. Update the nodal velocities: ${}^{n+1}\boldsymbol{\bar{v}}^{i+1} = {}^{n+1}\boldsymbol{\bar{v}}^i + \boldsymbol{\Delta\bar{v}}$

3. Update the nodal coordinates: ${}^{n+1}\boldsymbol{\bar{x}}^{i+1} = {}^{n+1}\boldsymbol{\bar{x}}^i + \boldsymbol{\bar{u}}(\boldsymbol{\Delta\bar{v}})$

4. Compute the updated stress tensors:

$${}^{n+1}\boldsymbol{\sigma}^{\nabla,i+1} = \boldsymbol{c}^{\sigma J} : \boldsymbol{d}\left(\boldsymbol{\bar{v}}^{i+1}\right) \quad \rightarrow \quad {}^{n+1}\boldsymbol{\sigma}^{i+1} = {}^n\boldsymbol{\hat{\sigma}} + \Delta t \,{}^{n+1}\boldsymbol{\sigma}^{\nabla,i+1}$$

5. Check convergence: $\parallel {}^{n+1}\boldsymbol{R}^{i+1}({}^{n+1}\boldsymbol{\bar{v}}^{i+1}) \parallel < tolerance$

If condition 5 is not fulfilled, return to 1 with $i \leftarrow i + 1$.
At the end of each time step: ${}^{n+1}\boldsymbol{\hat{\sigma}} = {}^{n+1}\boldsymbol{\sigma} + \Delta t \boldsymbol{\Omega}\left({}^{n+1}\boldsymbol{\bar{v}}, {}^{n+1}\boldsymbol{\sigma}\right)$

Box 3. Iterative solution scheme for hypoelastic solids using the velocity formulation

All the matrices and vectors in Box 3 are collected in Box 4

$$K_{IJ}^\rho = I \int_\Omega N_I \frac{2\rho}{\Delta t} N_J d\Omega \,, \quad K_{IJ}^g = I \int_\Omega \beta_I^T \Delta t \sigma \beta_J d\Omega$$

$$K_{IJ}^m = \int_\Omega \boldsymbol{B}_I^T \Delta t \left[\boldsymbol{c}^{\sigma J}\right] \boldsymbol{B}_J d\Omega$$

$$\boldsymbol{c}^{\sigma J} = \begin{bmatrix} \kappa + \frac{4}{3}\mu & \kappa - \frac{2}{3}\mu & \kappa - \frac{2}{3}\mu & 0 & 0 & 0 \\ \kappa - \frac{2}{3}\mu & \kappa + \frac{4}{3}\mu & \kappa - \frac{2}{3}\mu & 0 & 0 & 0 \\ \kappa - \frac{2}{3}\mu & \kappa - \frac{2}{3}\mu & \kappa + \frac{4}{3}\mu & 0 & 0 & 0 \\ 0 & 0 & 0 & \mu & 0 & 0 \\ 0 & 0 & 0 & 0 & \mu & 0 \\ 0 & 0 & 0 & 0 & 0 & \mu \end{bmatrix}$$

$$R_{Ii} = \int_\Omega N_I \rho N_J d\Omega \, \bar{v}_{Ji} + \int_\Omega \frac{\partial N_I}{\partial x_j} \sigma_{ij} d\Omega - \int_\Omega N_I b_i d\Omega - \int_{\Gamma_t} N_I t_i^p d\Gamma$$

where $\kappa = \left(\frac{2\mu}{3} + \lambda\right)$ and λ and μ are the Lamé constants

Box 4. Element form of the matrices and vectors in Box 3

2.3.2   Mixed Velocity–Pressure Formulation for Hypoelastic Solids

The solution scheme is like the one presented in Sect. 2.2.2. As already explained, the tangent matrix of the mixed Velocity–Pressure formulation is the same as for the Velocity formulation. However the Cauchy stress tensor is computed from the nodal velocities and the nodal pressures using Eq. (2.99). The governing equations are Eqs. (2.70)–(2.71) particularized with the material parameters of a hypoelastic solid. The finite element implemented with this hypoelastic mixed Velocity–Pressure formulation is called VP-element.

In Box 5, the iterative solution incremental scheme for hypoelastic solids using the mixed Velocity–Pressure formulation is given for a generic time interval $[^n t, {}^{n+1} t]$.

For each iteration i within a time interval:

1. Compute the nodal velocity increments $\Delta \bar{v}$:

$$K^i \Delta \bar{v} = R^i ({}^{n+1} \bar{v}^i, {}^{n+1} \bar{p}^i)$$

   where $K^i = K^m ({}^{n+1} \bar{x}^i, c^{\sigma J}) + K^g ({}^{n+1} \bar{x}^i, \sigma^i (\bar{v}^i, \bar{p}^i)) + K^\rho ({}^{n+1} \bar{x}^i)$

2. Update the nodal velocities: ${}^{n+1} \bar{v}^{i+1} = {}^{n+1} \bar{v}^i + \Delta \bar{v}$

3. Update the nodal coordinates: ${}^{n+1} \bar{x}^{i+1} = {}^{n+1} \bar{x}^i + \bar{u}(\Delta \bar{v})$

4. Compute the nodal pressures ${}^{n+1} \bar{p}^{i+1}$:

$$\frac{1}{\Delta t} M_1 {}^{n+1} \bar{p}^{i+1} = Q^T {}^{n+1} \bar{v}^{i+1} + g^n ({}^n \bar{p}^i)$$

5. Compute the updated stress measures:

$${}^{n+1} \sigma^{i+1} = {}^n \hat{\sigma} + {}^{n+1} \Delta \bar{p}^{i+1} I + 2\mu \Delta t \left[I' : d \left(\bar{v}^{i+1} \right) \right]$$

6. Check convergence: $\| {}^{n+1} R^{i+1} ({}^{n+1} v^{i+1}, {}^{n+1} p^{i+1}) \| < tolerance$

If condition 6 is not fulfilled, return to 1 with $i \leftarrow i + 1$.

At the end of each time step: ${}^{n+1} \hat{\sigma} = {}^{n+1} \sigma + \Delta t \Omega \, ({}^{n+1} \bar{v}, {}^{n+1} \sigma)$

Box 5. Iterative solution scheme for hypoelastic solid using mixed Velocity–Pressure formulation

All the matrices and vectors that appear in Box 5 are collected in Box 6

Vectors and matrices for the linear momentum equations

$$K^\rho_{IJ} = I \int_\Omega N_I \frac{2\rho}{\Delta t} N_J d\Omega \,, \quad K^g_{IJ} = I \int_\Omega \beta^T_I \Delta t \sigma \beta_J d\Omega$$

$$K^m_{IJ} = \int_\Omega B^T_I \Delta t \left[c^{\sigma J} \right] B_J d\Omega$$

$$c^{\sigma J} = \begin{bmatrix} \kappa + \frac{4}{3}\mu & \kappa - \frac{2}{3}\mu & \kappa - \frac{2}{3}\mu & 0 & 0 & 0 \\ \kappa - \frac{2}{3}\mu & \kappa + \frac{4}{3}\mu & \kappa - \frac{2}{3}\mu & 0 & 0 & 0 \\ \kappa - \frac{2}{3}\mu & \kappa - \frac{2}{3}\mu & \kappa + \frac{4}{3}\mu & 0 & 0 & 0 \\ 0 & 0 & 0 & \mu & 0 & 0 \\ 0 & 0 & 0 & 0 & \mu & 0 \\ 0 & 0 & 0 & 0 & 0 & \mu \end{bmatrix}$$

$$R_{Ii} = \int_\Omega N_I \rho N_J d\Omega \, \bar{v}_{Ji} + \int_\Omega \frac{\partial N_I}{\partial x_j} \sigma'_{ij} d\Omega +$$
$$+ \int_\Omega \frac{\partial N_I}{\partial x_j} \delta_{ij} N_J d\Omega \, \bar{p}_J - \int_\Omega N_I b_i d\Omega - \int_{\Gamma_t} N_I t^p_i d\Gamma$$

Vectors and matrices for the continuity equation

$$M_{1_{IJ}} = \int_\Omega N_I \frac{1}{\kappa} N_J d\Omega \,, \quad Q_{IJ} = \int_\Omega B^T_I m N_J d\Omega$$

$${}^n g_I = \int_\Omega N_I \frac{1}{\kappa \Delta t} N_J d\Omega \, {}^n \bar{p}_J$$

where $\kappa = \left(\frac{2\mu}{3} + \lambda \right)$ and λ and μ are the Lamé constants

---

Box 6. Element form of the matrices and vectors of Box 5

Note that for the mixed Velocity–Pressure formulation the material part of the tangent matrix is defined by the same tangent moduli of the velocity scheme ($c^{\sigma J}$).

In the mixed formulation, the momentum and the continuity equations can be easily decoupled. This is obtained by computing the Cauchy stress tensor using the velocities only (Eq. (2.98)) and not as the sum of its deviatoric part and the pressure (Eq. (2.99)). The uncoupled mixed Velocity–Pressure formulation is totally equivalent to the velocity formulation. In fact, although the pressures are still computed by solving the continuity equation, they do not appear in the momentum equations and, hence, they do not affect the solution for each time step.

2.3.3 Theory of Plasticity

The theory of plasticity is dedicated to those solids that, after being subjected to a loading process, may sustain permanent (*plastic*) deformations when completely unloaded [11]. The plasticity is defined *rate-independent* if the permanent deformations of the material do not depend on the rate of application of the loads. The materials whose behavior can be adequately described by this theory are called *rate-independent plastic* materials.

Elastic-plastic laws are characterized for being path-dependent and dissipative. The stresses cannot be computed as a single-valued function of the strains because they depend on the entire history of the deformation [1].

The most important properties of the theory of plasticity can be summarized as follows:

1. The increments of strain $d\varepsilon$ are assumed to be additively decomposed into an elastic (reversible) part $d\varepsilon_{el}$ and a plastic (irreversible) part $d\varepsilon_{pl}$, such that

$$d\varepsilon = d\varepsilon_{el} + d\varepsilon_{pl} \tag{2.100}$$

2. There exists an *elastic domain* where the behavior of the material is purely elastic and permanent deformations are not produced;
3. The *yield function* f_Y delimits the elastic domain. It governs the onset and the continuity of the plastic deformations and it is a functions of the state of stress and of the internal variables $\boldsymbol{q}$. So

$$f_Y = f_Y(\boldsymbol{\sigma}, \boldsymbol{q}) \tag{2.101}$$

The yield function cannot be positive: it is negative when the stress state is below the yield value and null otherwise (*yield condition*: $f_Y = 0$);
4. The plastic strain increments are governed by the so called *flow rule*;
5. $\dot{\lambda}_{pl}$ is the *plastic strain parameter* and it is positive for a plastic loading and equal to zero for elastic loading or unloading;
6. The loading-unloading process is described by the Khun–Tucker conditions

$$\dot{\lambda}_{pl} \geq 0, \quad f_Y \leq 0, \quad \dot{\lambda}_{pl} f_Y = 0 \tag{2.102}$$

The third condition can also be expressed in the rate form through the so-called *consistency condition*, $\dot{f}_Y = 0$. For plastic loading ($\dot{\lambda}_{pl} > 0$) the stress state lies on the yield surface ($f_Y = 0$), instead for elastic loading or unloading the yield condition is not reached ($f_Y < 0$) and there is not plastic flow ($\dot{\lambda}_{pl} = 0$).

2.3.3.1 Hypoelastic–Plastic Materials

Hypoelastic–plastic models are typically used when the elastic strains represent only a small part of the total strains. In other words, when the plastic strains are much larger than the elastic ones. This is because of the inaccuracy of the hypoelastic models in the large strain regime. However, if the elastic strains are small, the energy error introduced by the hypoelastic description of the elastic response is limited and can be considered adequate [1].

Depending on the problem, the yield function can be based on a different constitutive model. For example, for soil plasticity the Drucker–Prager model is the most used, while for porous plastic solids the Gurson model is more adequate. In this work, the J_2 von Mises flow model is used. This model is particularly indicated for the metal plasticity [1].

The hypoelastic–plastic model described in this section has been taken from [1]. According to the von Mises criterion [12], plastic yielding begins when the J_2 stress deviator invariant reaches a critical value. The J_2 stress deviator invariant is defined as

$$J_2 = \frac{1}{2}\sigma'_{ij}\sigma'_{ij} \tag{2.103}$$

The key assumption of the von Mises model is that the plastic flow is not affected by the pressure but only by the deviatoric stress. This hypothesis has been experimentally verified for metals [13]. For this reason the von Mises model is called to be *pressure insensitive*.

A yield function for the von Mises criterion can be defined as

$$f(\boldsymbol{\sigma}, \boldsymbol{q}) = \bar{\sigma} - \sigma_Y \tag{2.104}$$

where σ_Y is the uniaxial yield stress and it is related to the shear yield stress τ_Y as follows

$$\sigma_Y = \sqrt{3}\tau_Y \tag{2.105}$$

and in Eq. (2.104) $\bar{\sigma}$ is the *von Mises effective* or *equivalent stress* defined as

$$\bar{\sigma} = \sqrt{3J_2} \tag{2.106}$$

Concerning the deformation, the elastic-plastic decomposition described in Eq. (2.100) is rewritten in terms of rates as

$$\boldsymbol{d} = \boldsymbol{d}_{el} + \boldsymbol{d}_{pl} \tag{2.107}$$

where $\boldsymbol{d}_{el}$ and $\boldsymbol{d}_{pl}$ are the deformation rates associated to the elastic and plastic responses, respectively.

Combining Eq. (2.107) with (2.83), yields

$$\sigma^{\nabla J} = c_{el}^{\sigma J} : (d - d_{pl})$$

(2.108)

The rate of plastic deformations is given by

$$d_{pl} = \dot{\lambda}_{pl} r(\sigma, q)$$

(2.109)

where the plastic flow rate $\dot{\lambda}_{pl}$ is a scalar and $r(\sigma, q)$ represents the plastic flow direction.

Substituting Eq. (2.109) in (2.108), yields

$$\sigma^{\nabla J} = c_{el}^{\sigma J} : (d - \dot{\lambda}_{pl} r)$$

(2.110)

During plastic loading the plastic flow rate is positive and the state of stress remains on the yield surface $f_Y = 0$. This is consistent with the third Khun–Tucker condition $\dot{\lambda} f_Y = 0$. The consistency condition $\dot{f}_Y = 0$ has the same meaning. Using the chain rule on the consistency condition, yields

$$\dot{f}_Y = \frac{\partial f_Y}{\partial \sigma} : \dot{\sigma} + \frac{\partial f_Y}{\partial q} \cdot \dot{q} = 0$$

(2.111)

If the yield function depend on the invariant, the following relation holds [1, 10]

$$\frac{\partial f_Y}{\partial \sigma} : \dot{\sigma} = \frac{\partial f_Y}{\partial \sigma} : \sigma^{\nabla J}$$

(2.112)

Combining Eqs. (2.112) and (2.110) and substituting in Eq. (2.111), yields

$$\frac{\partial f_Y}{\partial \sigma} : c_{el}^{\sigma J} : (d - d_{pl}) + \frac{\partial f_Y}{\partial q} \cdot \dot{q} = 0$$

(2.113)

In the most plastic models the evolution of the function q of the internal variables h can be expressed as a function of the plastic strain parameter as follows

$$\dot{q} = \dot{\lambda} h$$

(2.114)

where h are the internal variables.

Substituting Eqs. (2.109) and (2.114) in (2.113), the following relation for the plastic strain parameter is obtained

$$\dot{\lambda} = \frac{\frac{\partial f_Y}{\partial \sigma} : c_{el}^{\sigma J} : d}{-\frac{\partial f_Y}{\partial q} \cdot h + \frac{\partial f_y}{\partial \sigma} : c_{el}^{\sigma J} : r}$$

(2.115)

The plastic flow vector r is often derived from a plastic flow potential. If the plastic flow potential coincides with the yield function, the plastic flow is termed *associative*. In this case r is proportional to the normal of the yield surface, that is $r \propto \frac{\partial f_Y}{\partial \sigma}$. An associative plastic flow has the important advantage that it can lead to a symmetric stiffness matrix [1]. In this work an associative plasticity and a constant plastic modulus H (for perfect plasticity, $H = 0$) have been considered. Using these hypotheses the plastic strain parameter is expressed as

$$\dot{\lambda} = \frac{r : c_{el}^{\sigma J} : d}{H + r : c_{el}^{\sigma J} : r} \tag{2.116}$$

Substituting this relation in Eq. (2.110), yields

$$\sigma^{\nabla J} = c_{el}^{\sigma J} : \left(d - \frac{r : c_{el}^{\sigma J} : d}{H + r : c_{el}^{\sigma J} : r} r \right) \tag{2.117}$$

The same can be computed using a tangent moduli over the whole deformation rate as

$$\sigma^{\nabla J} = c^{\sigma J} : d = \left[c_{el}^{\sigma J} - \frac{\left(c_{el}^{\sigma J} : r \right) \otimes \left(r : c_{el}^{\sigma J} \right)}{H + r : c_{el}^{\sigma J} : r} \right] : d \tag{2.118}$$

where $c^{\sigma J}$ is the *continuum* elasto-plastic tangent modulus.

For associative plasticity, the von Mises plastic flow is computed as

$$r = \frac{\partial f}{\partial \sigma} = \frac{3}{2\bar{\sigma}} \sigma' \tag{2.119}$$

Because the plastic flow vector r is deviatoric it follows that

$$c_{el}^{\sigma J} : r = 2\mu \quad , \quad r : c_{el}^{\tau J} : r = 3\mu \tag{2.120}$$

Form Eqs. (2.86), (2.118) and (2.120), the following expression of the elastoplastic modulus is obtained

$$c^{\sigma J} = \kappa I \otimes I + 2\mu I' - 2\mu\gamma n \otimes n \tag{2.121}$$

with

$$\gamma = \frac{1}{1 + (H/3\mu)} \tag{2.122}$$

$$n = \sqrt{\frac{2}{3}} r \tag{2.123}$$

For perfect plasticity $H = 0$, so $\gamma = 1$ and Eq. (2.121) simplifies to

$$c^{\sigma J} = \kappa I \otimes I + 2\mu I' - 2\mu n \otimes n \tag{2.124}$$

Note that the continuum elasto-plastic tangent modulus conserves the symmetry properties of its elastic counterpart. For elastic loading or unloading, $c^{\sigma J} = c_{el}^{\sigma J}$.

For a plane strain state, $c^{\sigma J}$ is

$$\left[c^{\sigma J}\right] = \begin{bmatrix} \kappa + \frac{4}{3}\mu & \kappa - \frac{2}{3}\mu & 0 \\ \kappa - \frac{2}{3}\mu & \kappa + \frac{4}{3}\mu & 0 \\ 0 & 0 & \mu \end{bmatrix} - 2\mu\gamma \begin{bmatrix} n_{xx}n_{xx} & n_{xx}n_{yy} & n_{xx}n_{xy} \\ n_{yy}n_{xx} & n_{yy}n_{yy} & n_{yy}n_{xy} \\ n_{xy}n_{xx} & n_{xy}n_{yy} & n_{xy}n_{xy} \end{bmatrix}$$

In order to guarantee the consistency of the elastoplastic incremental scheme, the so-called *return mapping* algorithm has to be introduced. With this technique, the Khun–Tucker conditions (Eq. (2.102)) are enforced at the end of a plastic time step in order to recover exactly the yield condition $f(\sigma_{n+1}) = 0$. The return mapping algorithm consists of an initial trial elastic step followed by a plastic corrector one that is activated when the yield function takes a positive value. In Fig. 2.1 from [14], a graphical representation of the return mapping algorithm is shown.

For associative plasticity, during the plastic corrector step driven by the increment of the plasticity parameter λ, the plastic flow direction r is normal to the yield surface.

For the J_2 flow theory and associative plasticity, the return mapping is characterized to be *radial* [15]. This because the von Mises yield surface is circular, thus its normal is also radial. In Fig. 2.2 a graphical representation of the radial return algorithm for J_2 plasticity is shown.

Fig. 2.1 Graphical representation of the return mapping algorithm [14]

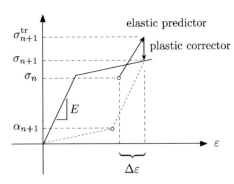

Fig. 2.2 Graphical representation of the radial return method for J_2 plasticity

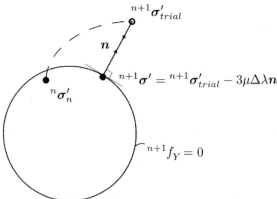

The return radial mapping procedure typically starts with the elastic prediction of the stresses. This means that the linear momentum equations are solved using only the elastic part of the continuum tangent modulus (Eq. (2.86)) and the stress tensor $\sigma^{(0)}$ is computed with Eq. (2.99) (the superindex refers to the iteration index).

Then the effective stress $\bar{\sigma}$ is computed with Eq. (2.106) and the yield function Eq. (2.104) is evaluated. If $f_Y > 0$ the return mapping iterative procedure is required. In fact, an elastoplastic step has been computed as purely elastic without fulfilling the consistency condition.

The first step of the iterative corrector procedure consists of computing the increment of the plastic parameter as

$$\delta\lambda_{pl}^{(k)} = \frac{f^{(k)}}{3\mu + H^{(k)}} \tag{2.125}$$

For perfect plasticity, the plastic modulus H is null, hence the plastic parameter is $\delta\lambda_{pl}^{(k)} = f^{(k)}/(3\mu)$.

The plastic parameter increment is updated as

$$\Delta\lambda_{pl}^{(k+1)} = \Delta\lambda_{pl}^{(k)} + \delta\lambda_{pl}^{(k)} \tag{2.126}$$

If before this step the material has never suffered plastic deformations, then $\Delta\lambda_{pl}^{(k)} = 0$.

Next the plastic strain and the internal variables are updated according to the plastic correction derived from Eq. (2.125).

The increment of plastic deformation is

$$\Delta\varepsilon_{pl}^{(k)} = -\delta\lambda_{pl}^{(k)}\sqrt{\frac{3}{2}}\boldsymbol{n} \tag{2.127}$$

So the total plastic deformations are

$$\varepsilon_{pl}^{(k+1)} = \varepsilon_{pl}^{(k)} + \Delta\varepsilon_{pl}^{(k)} \tag{2.128}$$

Once again, for the first plastic step $\varepsilon_{pl}^{(k)} = 0$.

Next the deviatoric stresses are updated as

$$\sigma'^{(k+1)} = \sigma'^{(0)} - 2\mu\Delta\lambda_{pl}^{(k+1)}\boldsymbol{r}^{(0)} \tag{2.129}$$

The plastic flow direction $\boldsymbol{r}$ remains unchanged because also the tensor $\boldsymbol{n}$ remains unchanged (and radial) during the plastic corrector phase (Eq. (2.123)). This is a particular feature of the radial return mapping.

From Eqs. (2.106) and (2.130), the updated effective stress is

$$\bar{\sigma}^{(k+1)} = \bar{\sigma}^{(0)} - 3\mu\Delta\lambda_{pl}^{(k+1)} \tag{2.130}$$

Then the yield condition (Eq. 2.104) is verified again with the updated effective stress. If it is not fulfilled, the steps from Eq. (2.125) to (2.130) are repeated until $f_Y^{(k)} = 0$.

In Box 5, the return mapping algorithm for the J_2 theory and referred to a general elastoplastic time interval $[^n t, \, ^{n+1} t]$ is summarized.

---

0. Initialization: $\varepsilon_{pl}^{(0)} = \, ^n \varepsilon_{pl}^{(0)}, \quad \Delta \lambda^{(0)} = 0;$

1. Elastic prediction:

$$\sigma^\nabla = C_{el}^\sigma : d \rightarrow Eq. \, (2.99) \rightarrow \sigma^{i+1} \left(\sigma^\nabla, \Omega \right)$$

2. Evaluation of the yield function $f^{(0)} = \bar{\sigma}^{(0)} - \sigma_Y$:

   $if \ f < tolerance \ \rightarrow$ go ahead; $else:$ go to loop of Step 3.

3. Plastic correction:

   3.0 Initialization: $k = 0, \quad \boldsymbol{n} = \sqrt{\dfrac{3}{2}} \dfrac{\sigma'}{\bar{\sigma}};$

   3.1 Compute the plasticity parameter increment:

   $$\delta \lambda_{pl}^{(k)} = \frac{f^{(k)}}{3\mu + H^{(k)}} \quad \rightarrow \quad \Delta \lambda_{pl}^{(k+1)} = \Delta \lambda_{pl}^{(k)} + \delta \lambda_{pl}^{(k)}$$

   3.2 Update the plastic strains:

   $$\Delta \varepsilon_{pl}^{(k)} = -\delta \lambda_{pl}^{(k)} \sqrt{\frac{3}{2}} \boldsymbol{n} \quad \rightarrow \quad \varepsilon_{pl}^{(k+1)} = \varepsilon_{pl}^{(k)} + \Delta \varepsilon_{pl}^{(k)}$$

   3.3 Update the stresses:

   $$\sigma'^{(k+1)} = \sigma'^{(0)} - 2\mu \Delta \lambda_{pl}^{(k+1)} \sqrt{\frac{3}{2}} \boldsymbol{n}$$

   $$\bar{\sigma}^{(k+1)} = \bar{\sigma}^{(0)} - 3\mu \Delta \lambda_{pl}^{(k+1)}$$

   3.4 Check the yield condition $f^{(k+1)} = \bar{\sigma}^{(k+1)} - \sigma_Y$:

   $if \ f^{(k+1)} < tolerance \ \rightarrow$ converged;

   $else: k \leftarrow k + 1$ and go to Step 3.1.

Box 5. Radial return mapping for J_2 plasticity

2.3.4 Validation Examples

In this section several problems are studied in order to validate and the V and the VP elements and to make interesting comparisons. First an example for small displacements is studied. Then a benchmark problem for non-linear solid mechanics, namely the Cook's membrane, is analyzed. The third example is a uniformly loaded circular plate and it involves also plasticity. All these examples are analyzed in statics considering a unique unit time step increment for the velocity-based formulations. The last example is solved in dynamics and for both the hypoelastic and hypoelastic–plastic models.

Simply Supported Beam

The first validated problem is a simply supported beam loaded by its self-weight. The problem has been studied in statics so the inertial forces have not been considered. The geometry of the problem is illustrated in Fig. 2.3 and the problem data are given in Table 2.1. The material properties of the structure can be assimilated to the ones of a structural steel.

The beam undergoes small displacements under the effect of its self-weight, hence linear elastic theory is suitable for computing a reference solution. The accuracy of the formulation is tested by comparing the computed values for the maximum vertical displacement and the maximum XX-component of the Cauchy stress tensor with the values given by a linear elastic analysis. According to this theory, both maximum values are reached in the central section of the beam and they are computed as

$$U_Y^{max} = \frac{5gHL^4}{384EI} = 1.5348 \cdot 10^{-4} \ m \tag{2.131}$$

$$\sigma_X^{max} = \frac{3gHL^2}{4H^2} = 2887818 \ Pa \tag{2.132}$$

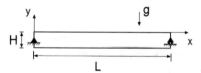

Fig. 2.3 Simply supported beam. Initial geometry

Table 2.1 Simply supported beam. Problem data

L	5 m
H	0.5 m
Young modulus	196 GPa
Density	$7.85 \cdot 10^3$ kg/m^3
Poisson ratio	0

(a) average mesh size= 0.25m

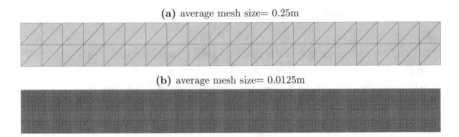

(b) average mesh size= 0.0125m

Fig. 2.4 Simply supported beam. Coarsest and finest meshes used for the analysis

The problem has been solved in 2D using both the Velocity and the mixed Velocity–Pressure formulations using 3-noded triangular elements. The static problem is solved with only one unit time increment.

In order to verify the convergence of both schemes, the problem has been solved using structured meshes of 3-noded triangles with the following average sizes: 0.25, 0.125, 0.05, 0.025, 0.0125 m. In Fig. 2.4 the finest (mesh size = 0.00125 m, 32,000 elements) and the coarsest (mesh size = 0.25m, 80 elements) meshes are illustrated.

In Fig. 2.5 the solutions for the vertical displacement and the XX-component of the Cauchy stress tensor computed at the Gauss points obtained with the mesh with average size 0.025 m are plotted.

For the visualization of all the numerical results of this work the pre-postprocessor software GID [16] has been used.

Table 2.2 collects the values of the maximum vertical displacement (absolute value) and the XX-component of the Cauchy stress tensor computed at the Gauss point using the V-element and the VP-element for different FEM mesh.

In the examples presented in this section, for the convergence analysis the percentage error is computed versus the solution obtained with the finest discretization as

$$error = abs \left(\frac{value_{finest\ mesh} - value_{tested\ mesh}}{value_{finest\ mesh}} \right) \cdot 100 \qquad (2.133)$$

For example, the maximum vertical displacement obtained with the finest mesh (average size 0.0125 m) is the reference solution for both V and VP elements. The convergence curves for both formulations are plotted in Fig. 2.6. Both elements show a quadratic convergence rate for this error measure.

Compressible Cook's Membrane

The Cook's membrane is a benchmark problem for solid mechanics. The static problem is solved twice in this thesis. In this section a compressible material is considered; in the next chapter the nearly incompressible case is analyzed. In both cases the problem has been solved with only one unit time increment.

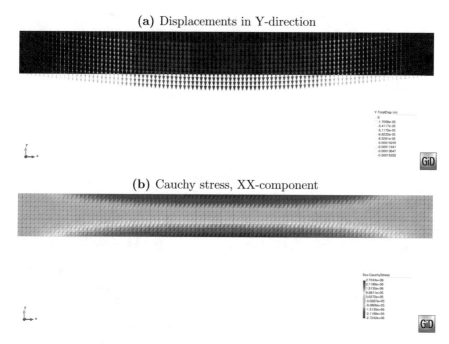

(a) Displacements in Y-direction

(b) Cauchy stress, XX-component

Fig. 2.5 Simply supported beam. Numerical results

Table 2.2 Simply supported beam. Cauchy stress XX-component and maximum vertical displacement for different discretizations

Mesh size	V-element		VP-element	
	σ_x^{max}	U_y^{max}	σ_x^{max}	U_y^{max}
2.50E-01	1.46E+06	9.92E-05	1.53E+06	8.41E-05
1.25E-01	2.29E+06	1.37E-04	2.37E+06	1.29E-04
5.00E-02	2.72E+06	1.54E-04	2.80E+06	1.52E-04
2.50E-02	2.82E+06	1.56E-04	2.87E+06	1.56E-04
1.25E-02	2.86E+06	1.57E-04	2.89E+06	1.57E-04

The initial geometry of the problem, as well the problem data are given in Fig. 2.7a.

The self weight of the membrane has not been taken into account in the analysis, so the membrane deforms under only the effect of the external load applied at its free edge. In this case the structure undergoes large displacements and the solution cannot been computed analytically. The results taken as the reference ones are those published in [17]. The comparison with the mentioned publication, as well the convergence test are performed for the vertical displacement of point A of Fig. 2.7a with coordinates $(x, y) = (48, 52)$. According to [17], under plane stress conditions this displacement is $U_Y^{max} = 23.964$.

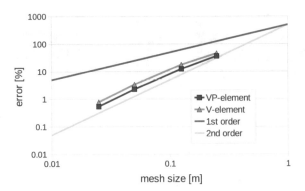

Fig. 2.6 Simply supported beam. Convergence analysis for the maximum vertical displacement for V and VP elements

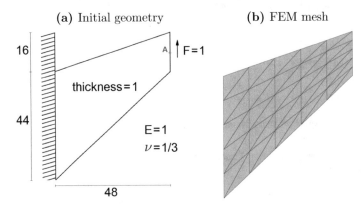

Fig. 2.7 Cook's membrane. Initial geometry, material data and FEM mesh (5 subdivisions for each edge corresponding to 50 elements)

The domain is discretized with a structured mesh and the edges of the membrane have the same number of partitions. In Fig. 2.7b one of the meshes used for this problem is given. The figure refers to the case of 5 elements for each edge of the membrane.

For the 2D simulation a convergence study has been performed using various discretizations. In the finest one the edges have 200 subdivisions, while in the coarsest one only 2.

The 3D problem (thickness = 1) has been solved for an unstructured mesh with average size 0.5 only. The results for this mesh are given in Fig. 2.8, where the XX-component of the Cauchy stress tensor is plotted over the deformed configuration.

For the 3D simulation, the vertical displacements at the central point of the free edge for the V and the VP elements are 23.942 and 23.952, respectively, which correspond to an error versus the reference solution of 0.092 % for the V-element and 0.050 % for the VP-element.

Fig. 2.8 Cook's membrane. Numerical results for the 3D simulation: the XX-component of the Cauchy stress tensor is plotted over the deformed configuration

The vertical displacement of point A of Fig. 2.7a obtained for all the 2D discretizations and for both the Velocity and the Velocity–Pressure formulations is plotted in the graph of Fig. 2.9.

Fig. 2.9 Cook's membrane. Vertical displacement of point A of Fig. 2.7a. Results for V and VP elements compared to the reference solution [17]

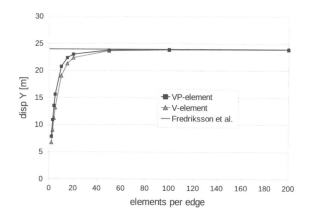

Table 2.3 Cook's membrane. Maximum vertical displacement for different discretizations

Elements per edge	V-element	VP-element
	U_y^{max}	U_y^{max}
2	6.707	7.8105
3	9.0274	10.901
4	11.232	13.515
5	13.1755	15.5985
10	19.037	20.729
15	21.272	22.332
20	22.349	22.987
50	23.658	23.781
100	23.878	23.913
200	23.941	23.95

In Table 2.3 the numerical values are given.

The convergence curves are given in Fig. 2.10. Also in this case the convergence rate is quadratic for both formulations.

Uniformly Loaded Circular Plate

The problem analyzed in this section is a simply supported circular plate subjected to a uniform pressure P on its top surface. The plate constrains are applied on its lower edge. The plate has a radius $R = 10$ and thickness $h = 1$. In this work, the axial symmetry of the problem has not been used, and the plate has been analyzed in 3D using 4-noded tetrahedra. The average size for the tetrahedra is 0.175. This gives 214,047 nodes. In Fig. 2.11 the FEM mesh used is shown.

A hypoelastic–perfectly plastic model has been used, and the problem has been solved with the mixed velocity pressure formulation. For the plastic part, a von Mises yield criterion has been considered. The plate has Young modulus $E = 10^7$, Poisson

Fig. 2.10 Cook's membrane. Convergence analysis for the vertical displacement of point A of Fig. 2.7a for the V and VP elements, V-element and VP-element, respectively

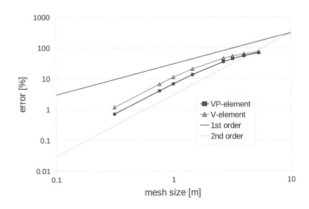

Fig. 2.11 Uniformly loaded circular plate. Initial geometry and 3D FEM used [18]

ratio $\nu = 0.24$ and a uniaxial yield stress $\bar{\sigma}_y = 16,000$. The objective of the study is to determine the limit load for the plate. Using the procedure described in [19], the limit load can be computed analytically by combining the limit analysis and the finite difference method. According to this theory, the limit load can be approximated as

$$P_{lim} \approx \frac{1.63\bar{\sigma}_y h^2}{R^2} = 260.8 \tag{2.134}$$

The same problem was solved in [11] using eight-noded axisymmetric quadrilateral elements with four Gauss integration points. The limit load obtained with a relatively coarse mesh (10 finite elements distributed in two layers across the thickness) is $P_{lim}^{FE} = 259.8$ [11].

As in [11], the limit load has been considered as the one for which the non-linear procedure cannot longer converge for a small increment of the load.

In Fig. 2.12 the maximum vertical displacement of the plate is plotted against the pressure on the top surface. In Table 2.4 the numerical values are given.

For the present analysis the limit load obtained is $P_{lim} = 264.27$, the relative percentage errors with respect the solutions given in [11, 19] are 1.37 % and 1.76 %, respectively.

In Fig. 2.13 the vertical displacements are depicted over the deformed configuration obtained with the limit load. The plate central section is highlighted in the picture.

In Fig. 2.14 some snapshots of the von Mises effective stress are plotted over the central section of the plate for the different load conditions. The picture shows clearly the progressive evolution of the plastic zone.

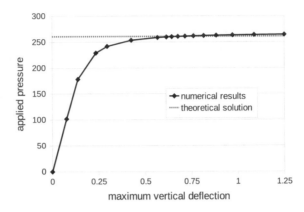

Fig. 2.12 Uniformly loaded circular plate. Maximum deflection versus the applied pressure [18]

Table 2.4 Uniformly loaded circular plate. Numerical values of the maximum vertical deflection
for different applied pressures

Pressure	Max. deflection	Pressure	Max. deflection
101.84	0.0758	260.71	0.677
178.22	0.138	261.21	0.716
229.14	0.236	261.73	0.761
241.87	0.296	262.24	0.816
253.58	0.424	262.73	0.885
258.67	0.567	263.26	0.972
259.69	0.615	263.77	1.088
260.20	0.644	264.27	1.250

Fig. 2.13 Uniformly loaded circular plate. Vertical displacement contours for the maximum pressure sustained by the plate ($P_{lim} = 264.27$) [18]

Plane Strain Cantilever in Dynamics

The plane strain cantilever illustrated in Fig. 2.15 has been chosen as the reference case for a large displacement dynamics analysis. The problem data are in given in Table 2.5. The problem was introduced and studied in [20]. In the reference publica-

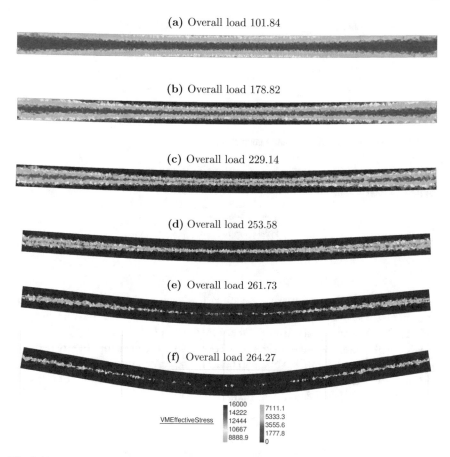

(a) Overall load 101.84

(b) Overall load 178.82

(c) Overall load 229.14

(d) Overall load 253.58

(e) Overall load 261.73

(f) Overall load 264.27

VMEffectiveStress

16000	7111.1
14222	5333.3
12444	3555.6
10667	1777.8
8888.9	0

Fig. 2.14 Uniformly loaded circular plate. Von Mises effective stress over the deformed configurations for different load conditions (only the central section is depicted) [18]

tion, the load was applied as a step function at time $t = 0$ s and its magnitude was defined by the equation $f = 15(1 - y^2/4)$ where y is the coordinate in the direction of the load. However, in this analysis the load has been considered uniformly distributed over the free edge (the overall value is 40, as in [20]).

The problem has been solved with both a hypoelastic and a hypoelastic–plastic models. First the results of the hypoelastic model are given.

Hypoelastic Model

The problem has been solved in 2D and 3D and using both the V and VP elements. In order to simulate the plane strain state, in the 3D analysis the nodal displacements in the transversal direction to the load have been constrained [20].

The reference solution is the elastic one given in [20].

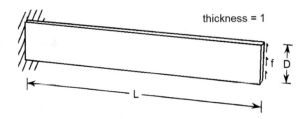

Fig. 2.15 Plane strain cantilever. Initial geometry

Table 2.5 Plane strain cantilever. Problem data

L	25
D	4
Young modulus	10^4
Poisson ratio	0.25

For the 2D analysis a convergence study has been performed. Structured finite element meshes have been used and the coarsest and the finest ones have a mean size of 1 and 0.125, respectively. Both meshes are given in Fig. 2.16.

For the 3D case, the problem has been solved with the finest mesh only (average size for the 4-noded tetrahedra equal to 0.125). The results for the 3D case obtained with the VP-element are illustrated in Fig. 2.17 where the pressure contours are plotted over the deformed configuration.

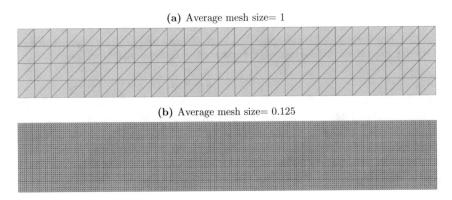

(a) Average mesh size= 1

(b) Average mesh size= 0.125

Fig. 2.16 Plane strain cantilever. Coarsest (mesh size = 1, 200 elements) and finest meshes (mesh size = 0.125, 12,800 *triangles*) used for the 2D analysis

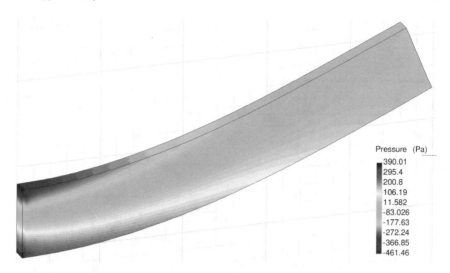

Fig. 2.17 Plane strain cantilever. Numerical results for the 3D simulation obtained with the VP-element: pressure contours plotted over the deformed configuration [18]

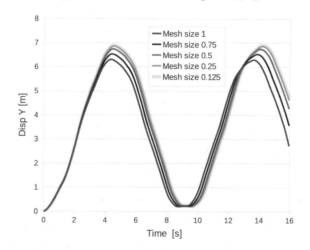

Fig. 2.18 Plane strain cantilever. Time evolution of the top corner vertical displacement for different 2D discretizations. Results obtained with the VP-element [18]

In Fig. 2.18 the time evolution of the top corner vertical displacement is plotted for each of the FEM meshes. These results have been obtained with the VP-element.

According to [20], the maximum vertical displacement is $U_Y^{max} = 6.88$. Table 2.6 collects the maximum vertical displacement obtained with the V and the VP elements for all the meshes.

Table 2.6 Plane strain cantilever. Maximum top corner vertical displacement for different 2D discretizations

Mesh size	V-element	VP-element
	U_y^{max}	U_y^{max}
1	5.759	6.306
0.8	6.144	6.534
0.5	6.568	6.743
0.25	6.811	6.863
0.125	6.875	6.895

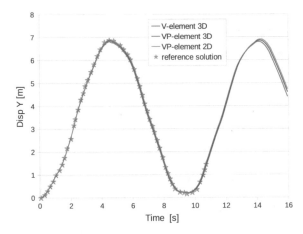

Fig. 2.19 Plane strain cantilever. Time evolution of the top corner vertical displacement. Solutions for the 2D VP-element and the 3D V and VP elements obtained with the finest mesh (average size 0.125) compared to the reference solution [18, 20]

The four curves of Fig. 2.19 are the converged time evolution of the top corner vertical displacement obtained with the V-element in 3D, the VP-element in 2D and 3D and the reference solution [20]. The curves corresponding to the V and VP elements are almost superposed and they match the reference solution.

Hypoelastic–plastic model

The same problem has been solved for an elastic-plastic material with linear hardening. The yield stress is 300 and the plastic modulus H is 100. The problem has been solved with the mixed Velocity–Pressure formulation and by using structured meshes, as the ones of Fig. 2.16. The reference solution is taken from [20] where the benchmark was proposed. In [20] the converged value for the maximum top corner vertical displacement is 8.22. The hypoelastic–plastic mixed Velocity–Pressure

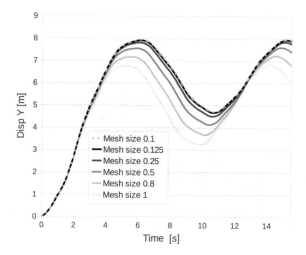

Fig. 2.20 Plane strain elastoplastic cantilever. Time evolution of the top corner vertical displacement for different 2D discretizations [18]

Table 2.7 Plane strain elastoplastic cantilever. Maximum and residual top corner vertical displacements for different discretizations

Mesh size	U_y^{max}	U_y^{res}
1	6.77	3.25
0.8	7.17	3.69
0.5	7.56	4.14
0.25	7.82	4.51
0.125	7.92	4.68
0.1	7.94	4.72
0.0625	7.97	4.77

formulation converges to 7.97 (error of 2.998 %). In the graph of Fig. 2.20 the time evolution of the top corner vertical displacement is plotted for the different FEM meshes.

In Table 2.7 the numerical values for the maximum and the residual top corner vertical displacements are given for each of the FEM mesh.

The problem has been solved also for the 3D problem for a structured mesh of 4-noded tetrahedra with average size 0.125.

In Fig. 2.21 the von Mises effective stresses are plotted over the deformed configuration at the time instant when the top corner vertical displacement is reached ($t = 6.05$ s).

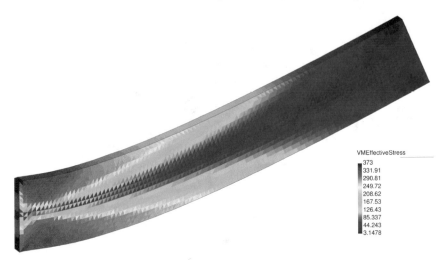

Fig. 2.21 Plane strain elastoplastic cantilever. Numerical results for the 3D simulation. Von Mises effective stress plotted over the deformed configuration at ($t = 6.05$ s) [18]

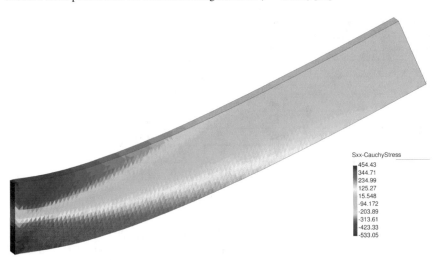

Fig. 2.22 Plane strain elastoplastic cantilever. Numerical results for the 3D simulation: XX-component of the Cauchy stress tensor plotted over the deformed configuration at ($t = 6.05$ s) [18]

In Fig. 2.22 for the same time instant the XX-component of the Cauchy stress tensor is plotted.

In Fig. 2.23 the 3D solution is compared to the 2D results obtained with a structured mesh with the same average size. The curves coincide almost exactly.

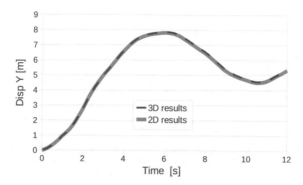

Fig. 2.23 Plane strain elastoplastic cantilever. Time evolution of the top corner vertical displacement. Numerical results for the 2D and the 3D simulations for the same average mesh size (0.125) [18]

2.4 Summary and Conclusions

In this chapter two velocity-based finite element Lagrangian procedures, namely the Velocity and the mixed Velocity–Pressure formulations, have been derived for a general compressible material.

The derivation of both formulations has been carried out with the aim of maintaining the scheme as general as possible. The mixed Velocity–Pressure formulation has been derived exploiting the linearized form of the Velocity formulation. In particular, it has been shown that the tangent matrix of the linear momentum equations is the same for both schemes.

The mixed Velocity–Pressure procedure is based on a two-step Gauss–Seidel solution algorithm. First, the linear momentum equations are solved for the velocity increments, next the continuity equation is solved for the pressure in the updated configuration. At the end of these steps the convergence for the velocities and the pressure is checked. Linear interpolation has been used for both velocity and pressure fields.

Next, both formulations have been particularized for hypoelastic solids. The finite elements generated from the Velocity formulation and the mixed Velocity–Pressure formulations have been called V and VP element, respectively.

The numerical scheme for dealing with J_2 associative plasticity has been also given.

Several numerical examples have been given for validating the V and VP elements for large displacements dynamics problems involving both hypoelastic and hypoelastoplastic compressible solids. It has been shown that both elements are convergent for all the numerical examples analyzed.

References

1. T. Belytschko, W.K. Liu, B. Moran, and K.I. Elkhodadry. *Nonlinear Finite Elements For Continua And Structures. Second Edition.* John Wiley & Sons, New York, 2014.
2. O.C. Zienkiewicz and R.L. Taylor. *The Finite Element Method. Its Basis and Fundamentals. (6th Ed.).* Elsevier Butterworth-Heinemann, Oxford, 2005.
3. F. Brezzi. On the existence, uniqueness and approximation of saddle-point problems arising from lagrange multipliers. *Revue française d'automatique, informatique, recherche opérationnelle. Série rouge. Analyse numérique,* 8 (R-2):129–151, 1974.
4. C. Truesdell. Hypo-elasticity. *Journal of Rational Mechanics and Analysis,* 4,1, 1955.
5. C. Truesdell. Remarks on hypo-elasticity. *Journal of research of the National Bureau of Standards - B. Mathematics and Mathematical Physics,* 67B (3):141–143, 1963.
6. C. Truesdell. The simplest rate theory of pure elasticity. *Communications on Pure and Applied Mathematics,* 8:123–132, 1955.
7. G.A. Holzapfel. *Nonlinear Solid Mechanics. A continuum Approach for Engineering.* John Wiley & Sons, New York, 2000.
8. C. Truesdell and W. Noll. *The Non-Linear Field Theories of Mechanics, Volume III.* Springer, New York, 2004.
9. J.C. Simo and T.J.R. Hughes. *Computational Inelasticity.* Springer, New York, 1998.
10. W. Prager. *Introduction to Mechanics of Continua.* Ginn and Company, Boston, 1961.
11. E.A. De Souza Neto, D. Peric, and D.R.J. Owen. *Computational methods for plasticity. Theory and applications.* John Wiley & Sons, New York, 2008.
12. R. von Mises. Mechanik der festen körper im plastisch- deformablen zustand. *Nachrichten von der Gesellschaft der Wissenschaften zu Göttingen, Mathematisch-Physikalische Klasse,* pages 582–592, 1913.
13. P. Bridgmain. *The Physics of High Pressure.* Bell & Sons, London, 1949.
14. R.I. Borja. *Plasticity. Modeling & Computation.* Springer, New York, 2013.
15. J.C. Simo and R.L. Taylor. Consistent tangent operators for rate-independent elastoplasticity. *Computer Methods in Applied Mechanics and Engineering,* 48:101–118, 1985.
16. GiD website. http://www.gidhome.com.
17. M. Fredriksson and N.S. Ottosen. Fast and accurate four-node quadrilateral. *International Journal For Numerical Methods In Engineering,* 61:1809–1834, 2004.
18. A. Franci, E. Oñate, and J. M. Carbonell. Velocity-based formulations for standard and quasi-incompressible hypoelastic-plastic solids. *International Journal for Numerical Methods in Engineering,* doi:10.1002/nme.5205., 2016.
19. J.J. Skrzypek. Plasticity and creep. theory, examples, and problems. *CRC Press,* London, 1993.
20. T. Belytschko and L.P. Bindeman. Assumed strain stabilization of eight node hexahedral element. *Computer Methods In Applied Mechanics And Engineering,* 105:225–260, 1993.

Chapter 3
Unified Stabilized Formulation for Quasi-incompressible Materials

This chapter is devoted to the derivation and validation of the unified stabilized formulation for nearly-incompressible materials. Namely, the cases of quasi-incompressible Newtonian fluids and quasi-incompressible hypoelastic solids will be analyzed.

Quasi-incompressible materials have a compressibility that is small enough to neglect the variation of density on time but, unlike fully incompressible materials, they are not totally divergence-free and the volumetric strain rate is related to the variation on time of pressure via Eq. (2.60). This stabilized formulation is based on the mixed Velocity–Pressure formulation derived in the previous chapter for a general material. In fact a one-field method, as the Velocity formulation presented in Chap. 2, is not sufficient for dealing with the incompressibility constraint. Furthermore the $inf-sup$ condition [1] imposes the stabilization of the mixed finite element procedure, if an equal order interpolation is used for the velocities and the pressure, as in this work. Consequently, the mixed Velocity–Pressure formulation derived for compressible materials in the previous chapter needs to be stabilized in order to solve quasi-incompressible problems.

In this work a new stabilized Lagrangian method for quasi-incompressible materials is derived. The stabilization procedure is based on the consistent derivation of a residual-based stabilized expression of the mass balance equation using the *Finite Calculus (FIC)*, also called *Finite Increment Calculus* method [2–8]. The main ideas of this part of the thesis are taken from [9] where the stabilization technique was derived for homogeneous viscous fluids. In this chapter it is shown that the procedure can be extended also for the analysis of quasi-incompressible solids.

The FIC approach in mechanics is based on expressing the equations of balance of mass and momentum in a space-time domain of finite size and retaining higher order terms in the Taylor series expansion typically used for expressing the change in the transported variables within the balance domain. In addition to the standard terms of infinitesimal theory, the FIC form of the balance equations contains derivatives of the classical differential equations in mechanics multiplied by characteristic distances in space and time.

© Springer International Publishing AG 2017
A. Franci, *Unified Lagrangian Formulation for Fluid and Solid Mechanics,*
Fluid-Structure Interaction and Coupled Thermal Problems Using the PFEM,
Springer Theses, DOI 10.1007/978-3-319-45662-1_3

In this work the second order FIC form in space and the first order FIC form in time of the mass balance equation have been used as the basis for the derivation of the stabilized formulation. The discretized variational form of the FIC mass balance equation via the FEM introduces terms in the Neumann boundary of the domain and other terms involving the first and second material time derivatives of the pressure. These terms are relevant for ensuring the consistency of the residual formulation.

The FIC stabilization, although is derived using the linear momentum equations, affects only the continuity equation. This means that the general form of the discretized and linearized momentum equations derived in Chap. 2 for the mixed Velocity–Pressure formulation still holds for quasi-incompressible materials. Hence, for hypoelastic quasi-incompressible materials the linear momentum equations are solved through the same linear system derived for the VP (compressible) element in Sect. 2.3.2. To avoid repetitions, the linearized momentum equations for the mixed Velocity–Pressure formulation will be recalled only for Newtonian fluids by deriving the tangent matrix and the internal forces according to the constitutive law.

For convenience, the stabilized form of the continuity equation is derived for quasi-incompressible Newtonian fluids, as in [9]. Nevertheless it will be shown that the approach can be easily extended to quasi-incompressible hypoelastic solids.

For the fluid analysis a Lagrangian procedure called Particle Finite Element Method (PFEM) [10] is used. With the PFEM, the mesh nodes are treated as particles and they move according to the governing equations. The domain is continuously remeshed using a procedure that efficiently combines the Delaunay tessellation and the Alpha Shape Method [11].

The FIC stabilized formulation here presented [9] has excellent mass preservation feature in the analysis of free surface fluid problems. Preservation of mass is a great challenge in the numerical study of flow problems with high values of the bulk modulus that approach the conditions of incompressibility. Mass losses can be induced by the stabilization terms which are typically added to the discretized form of the momentum and mass balance equations in order to account for high convective effects in the Eulerian description of the flow, and to satisfy the $inf-sup$ condition imposed by the full incompressibility constraint when equal order interpolation of the velocities and the pressure is used in mixed FEMs [12–15].

An important source of mass loss emanates in the numerical solution of free surface flows due, among other reasons, to the inaccuracies in predicting the shape of the free surface during large flow motions [16]. Mass losses can also occur in the numerical solution of flows with heterogeneous material properties [17] and in homogeneous viscous flows using the Laplace form of the Navier–Stokes equations [18].

In Lagrangian analysis procedures (such as the PFEM) the motion of the fluid particles is tracked during the transient solution. Hence, the convective terms vanish in the momentum equations and no numerical stabilization is needed for treating those terms. Two other sources of mass loss, however, remain in the numerical solution of Lagrangian flows, i.e. that due to the treatment of the incompressibility constraint by a stabilized numerical method, and that induced by the inaccuracies in tracking the flow particles and, in particular, the free surface.

The discretized variational form of the FIC mass balance equation via the FEM introduces terms in the Neumann boundary of the domain, and other terms involving the first and second material time derivatives of the pressure that are relevant for ensuring the consistency of the residual formulation. These terms are also crucial for preserving the mass during the transient solution of free surface Lagrangian flows. In addition they enable the computation of the nodal pressures from the stabilized mass balance equation without imposing any condition on the pressure at the free surface nodes, thus eliminating another source of mass loss which occurs when the pressure is prescribed to a zero value on the free surface in viscous flows.

A section of this chapter is exclusively dedicated to show the excellent mass preservation features of the PFEM-FIC stabilized formulation for free surface flow problems.

Various approaches have been developed in the recent years for approximating fluid flows by means of a quasi-incompressible material. In practice, they all consider the Navier–Stokes problem with a modified mass conservation equation where a slight compressibility is added to the fluid. In previous works, see for example [7, 15, 19, 20], the fluid compressibility was introduced by relaxing the incompressibility constraint by means of a penalty parameter. Alternatively, the small compressibility of the fluid can be introduced by considering the actual bulk modulus of the fluid κ, which gives this operation a physical meaning, [13, 15, 21]. The success of quasi-incompressible formulations in fluid mechanics relies on their important advantages from the numerical point of view. The most obvious one is that the quasi-incompressible form of the continuity equation yields a direct relation between the two unknown fields of the Navier–Stokes problem, the velocities and the pressure (Eq. (2.60)). This is useful if the problem is solved using a partitioned scheme because the Velocity–Pressure relation is crucial for deriving the tangent matrix of the momentum equations. Furthermore, another important drawback of fully incompressible schemes is eluded; the incompressibility constraint leads to a diagonal block of a zero matrix in the global matrix system. Consequently, a pivoting procedure is required to solve numerically this kind of linear system. It is well known that the computational cost associated to this operation is high and it increases with the number of degrees of freedom of the problem. The compressibility terms that emanate from the quasi-incompressible form of the continuity equation fill the diagonal of the global matrix, thus overcoming these numerical difficulties.

On the other hand, quasi-incompressible schemes insert in the numerical model parameters that have typically high values and can lead to different numerical instabilities. For example, large values of the penalty parameter or, equally, physical values of the bulk modulus, can compromise the quality of the analyses, or even prevent the convergence of the solution scheme, [15]. For this reason, generally, the value of the actual bulk modulus is reduced arbitrarily to the so-called *pseudo bulk modulus*. Nevertheless, an excessively small value of the pseudo bulk modulus changes drastically the meaning of the continuity equation of the original Navier–Stokes problem. In other words, the incompressibility constraint would not be satisfied at all. Furthermore, the bulk modulus is proportional to the speed of sound propagating through the material. Hence, we have to guarantee that the order of magnitude of the velocities of the problem is several times smaller than the velocity of sound in the medium.

In this work, the pseudo-bulk modulus κ_p is used for the tangent matrix of the linear momentum equations, while the actual physical value of the bulk modulus κ is used for the numerical solution of the mass conservation equation. The pseudo bulk modulus is computed as a proportion of the real bulk modulus of the fluid through the parameter θ such that $\kappa_p = \theta\kappa$ with $0 < \theta \leq 1$. A new numerical strategy for computing a priori the optimum value for the pseudo bulk modulus is also derived. For free surface flow problems, it will be shown that the scheme guarantees the good conditioning of the linear system, good convergence and excellent mass preservation features.

The lay-out of this chapter is the following. First the stabilized FIC form of the mass balance equation is derived. The procedure is described for a Newtonian fluid from the local and continuous form until the fully discretized matrix form. Next the linearized and discretized expression of the linear momentum equations derived in Chap. 2 for the mixed Velocity–Pressure formulation is adapted for Newtonian fluids and the complete solution scheme is given. Next the FIC stabilization procedure is extended to hypoelastic quasi-incompressible materials and the complete solution scheme is given also for this constitutive model.

In Sect. 3.4, free surface flows are analyzed in detail. First the essential features of the PFEM are given, then the mass preservation feature of the PFEM-FIC stabilized formulation is shown together with some explicative numerical examples. Finally the conditioning of the linear system is studied and a practical and efficient technique to predict a priori the optimum value for the pseudo bulk modulus is given.

The chapter ends up with several validation examples for quasi-incompressible solid and fluid mechanics problems.

3.1   Stabilized FIC Form of the Mass Balance Equation

The FIC stabilized form of the continuity equation is here derived. For convenience, the derivation procedure is carried out for quasi-incompressible Newtonian fluids.

3.1.1   Governing Equations

For the sake of clarity, the local forms of the linear momentum r_{mi} and the continuity r_v equations are recalled. From Eqs. (2.1) and (2.60) yields

$$r_{mi} := \rho \frac{\partial v_i}{\partial t} - \frac{\partial \sigma_{ij}}{\partial x_j} - b_i = 0 \quad , \quad i, j = 1, n_s \quad \text{in } \Omega \tag{3.1}$$

$$r_v := -\frac{1}{\kappa}\frac{\partial p}{\partial t} + d^v = 0 \tag{3.2}$$

The terms of Eqs. (3.1) and (3.2) have already been defined in the previous chapters. The standard form of the constitutive relation for a Newtonian fluid reads

$$\sigma_{ij} = \sigma'_{ij} + p\delta_{ij} = 2\mu d'_{ij} + p\delta_{ij} \tag{3.3}$$

where μ is the viscosity and the deviatoric strain rate d'_{ij} is defined from Eq. (2.56) as

$$d'_{ij} = d_{ij} - \frac{1}{3}d^v\delta_{ij} \tag{3.4}$$

where d^v is the volumetric strain rate.

Substituting Eqs. (3.3) and (3.4) into (3.1), gives a useful form of the momentum equations for the Newtonian fluids as

$$\rho\frac{\partial v_i}{\partial t} - \frac{\partial}{\partial x_j}(2\mu d_{ij}) + \frac{\partial}{\partial x_i}\left(\frac{2}{3}\mu d^v\right) - \frac{\partial p}{\partial x_i} - b_i = 0 \quad , \quad i,j = 1, n_s \tag{3.5}$$

3.1.2 FIC Mass Balance Equation in Space and in Time

Previous stabilized FEM formulations for quasi and fully incompressible fluids and solids were based on the first order form of the Finite Calculus (FIC) balance equation in space [3, 5–7, 9, 22–29]. In this work, for the derivation of stabilized formulation both the second order FIC form of the mass balance equation in space [7, 8] and the first order FIC form of the mass balance equation in time are used. These forms read respectively [9]

$$r_v + \frac{h_i^2}{12}\frac{\partial^2 r_v}{\partial x_i^2} = 0 \quad \text{in } \Omega \quad i = 1, n_s \tag{3.6}$$

and

$$r_v + \frac{\delta}{2}\frac{\partial r_v}{\partial t} = 0 \quad \text{in } \Omega \tag{3.7}$$

Equation (3.6) is obtained by expressing the balance of mass in a rectangular domain of finite size with dimensions $h_1 \times h_2$ (for 2D problems), where h_i are arbitrary distances, and retaining up to third order terms in the Taylor series expansions used for expressing the change of mass within the balance domain. The derivation of Eq. (3.6) for 2D incompressible flows can be found in [8].

Equation (3.7), on the other hand, is obtained by expressing the balance of mass in a space-time domain of infinitesimal length in space and finite dimension δ in time [3].

The FIC terms in Eqs. (3.6) and (3.7) play the role of space and time stabilization terms respectively. In the discretized problem, the space dimensions h_i and the time dimension δ are related to characteristic element dimensions and the time step increment, respectively as it will be explained later.

Note that for $h_i \to 0$ and $\delta \to 0$ the standard form of the mass balance equation (3.2), as given by the infinitesimal theory, is recovered.

3.1.3   FIC Stabilized Local Form of the Mass Balance Equation

Substituting Eq. (3.2) into (3.6) and (3.7) the second order FIC form in space and the first order FIC form in time of the mass balance equation for a general quasi-incompressible material read

$$-\frac{1}{\kappa}\frac{\partial p}{\partial t} + d^v - \frac{h_i^2}{12}\frac{\partial^2}{\partial x_i^2}\left(\frac{1}{\kappa}\frac{\partial p}{\partial t}\right) + \frac{h_i^2}{12}\frac{\partial}{\partial x_i}\left(\frac{\partial d^v}{\partial x_i}\right) = 0 \quad \text{in } \Omega \quad i = 1, n_s \tag{3.8}$$

and

$$-\frac{1}{\kappa}\frac{\partial p}{\partial t} + d^v - \frac{\delta}{2\kappa}\frac{\partial^2 p}{\partial t^2} + \frac{\delta}{2}\frac{\partial d^v}{\partial t} = 0 \tag{3.9}$$

The FIC form of the mass balance equation is expressed in terms of the momentum equations. Neglecting the space changes of the viscosity μ, from Eq. (3.5) the following expression is obtained

$$\frac{2}{3}\mu\frac{\partial d^v}{\partial x_i} = -\rho\frac{\partial v_i}{\partial t} + 2\mu\frac{\partial}{\partial x_j}(d_{ij}) + \frac{\partial p}{\partial x_i} + b_i = -\rho\frac{\partial v_i}{\partial t} + \hat{r}_{m_i} \tag{3.10}$$

Hence

$$\frac{\partial d^v}{\partial x_i} = \frac{3}{2\mu}\left[-\rho\frac{\partial v_i}{\partial t} + \hat{r}_{m_i}\right] \tag{3.11}$$

In the above two equations $\hat{r}_{m_i}$ is a *static momentum term* defined as

$$\hat{r}_{m_i} = 2\mu\frac{\partial}{\partial x_j}(d_{ij}) + \frac{\partial p}{\partial x_i} + b_i \tag{3.12}$$

Substituting Eq. (3.11) into (3.8) and neglecting the space changes of c and ρ in the derivatives, the following form is obtained

$$-\frac{1}{\kappa}\frac{\partial p}{\partial t} + d^v - \frac{h_i^2}{12}\frac{\partial^2}{\partial x_i^2}\left(\frac{1}{\kappa}\frac{\partial p}{\partial t}\right) + \frac{h_i^2}{8\mu}\frac{\partial}{\partial x_i}\left(-\rho\frac{\partial v_i}{\partial t} + \hat{r}_{m_i}\right) = 0 \tag{3.13}$$

Observation of the term involving the material derivative of v_i in Eq. (3.13) gives

$$\frac{\partial}{\partial x_i}\left(-\rho\frac{\partial v_i}{\partial t}\right) = -\rho\frac{\partial}{\partial t}\left(\frac{\partial v_i}{\partial x_i}\right) = -\rho\frac{\partial d^v}{\partial t} \tag{3.14}$$

Substituting Eq. (3.14) into (3.13) gives

$$-\frac{1}{\kappa}\frac{\partial p}{\partial t} + d^v - \frac{h_i^2}{12\kappa}\frac{\partial^2}{\partial x_i^2}\left(\frac{\partial p}{\partial t}\right) + \frac{h_i^2}{8\mu}\left(-\rho\frac{\partial d^v}{\partial t} + \frac{\partial \hat{r}_{m_i}}{\partial x_i}\right) = 0 \qquad (3.15)$$

From Eq. (3.9),

$$-\frac{\partial d^v}{\partial t} = -\frac{2}{\delta\kappa}\frac{\partial p}{\partial t} + \frac{2}{\delta}d^v - \frac{1}{\kappa}\frac{\partial^2 p}{\partial t^2} \qquad (3.16)$$

Substituting Eq. (3.16) into (3.15) gives

$$-\frac{1}{\kappa}\frac{\partial p}{\partial t} + d^v - \frac{h_i^2}{12\kappa}\frac{\partial^2}{\partial x_i^2}\left(\frac{\partial p}{\partial t}\right) + \frac{h_i^2}{8\mu}\left(-\frac{2\rho}{\delta\kappa}\frac{\partial p}{\partial t} + \frac{2\rho}{\delta}d^v - \frac{\rho}{\kappa}\frac{\partial^2 p}{\partial t^2} + \frac{\partial \hat{r}_{m_i}}{\partial x_i}\right) = 0$$

$$\qquad (3.17)$$

Multiplying Eq. (3.17) by $\frac{8\mu}{h^2}$ gives, after grouping some terms,

$$-\frac{1}{\kappa}\frac{\partial p}{\partial t}\left(\frac{8\mu}{h_i^2} + \frac{2\rho}{\delta}\right) + d^v\left(\frac{8\mu}{h_i^2} + \frac{2\rho}{\delta}\right) - \frac{2\mu}{3\kappa}\frac{\partial^2}{\partial x_i^2}\left(\frac{\partial p}{\partial t}\right) - \frac{\rho}{\kappa}\frac{\partial^2 p}{\partial t^2} + \frac{\partial \hat{r}_{m_i}}{\partial x_i} = 0$$

$$\qquad (3.18)$$

After some further transformations,

$$-\frac{1}{\kappa}\frac{\partial p}{\partial t} + d^v - \frac{2\mu\tau}{3\kappa}\frac{\partial}{\partial x_i}\left(\frac{\partial}{\partial x_i}\left(\frac{\partial p}{\partial t}\right)\right) - \tau\frac{\rho}{\kappa}\frac{\partial^2 p}{\partial t^2} + \tau\frac{\partial \hat{r}_{m_i}}{\partial x_i} = 0 \qquad (3.19)$$

where τ is a *stabilization parameter* given by

$$\tau = \left(\frac{8\mu}{h^2} + \frac{2\rho}{\delta}\right)^{-1} \qquad (3.20)$$

For *transient problems* the stabilization parameter τ is computed as

$$\tau = \left(\frac{8\mu}{(l^e)^2} + \frac{2\rho}{\Delta t}\right)^{-1} \qquad (3.21)$$

where Δt is the time step used for the transient solution and l^e is a characteristic element length.

The coefficient $\frac{2\mu\tau}{3\kappa}$ multiplying the second space derivatives of $\frac{\partial p}{\partial t}$ in Eq. (3.19) is much smaller than the coefficients multiplying the rest of the terms in this equation. Numerical tests have shown that the results are not affected by this term. Consequently, this second space derivative term will be neglected in the rest of this work.

Hence the FIC stabilized form of the mass balance equation is written as

$$-\frac{1}{\kappa}\frac{\partial p}{\partial t} + d^v - \tau\frac{\rho}{\kappa}\frac{\partial^2 p}{\partial t^2} + \tau\frac{\partial \hat{r}_{m_i}}{\partial x_i} = 0 \qquad (3.22)$$

Note that the term $\frac{\partial}{\partial x_i}\left(2\mu\frac{\partial}{\partial x_j}(d_{ij})\right)$ within $\hat{r}_{m_i}$ in Eq. (3.22) (see the definition of $\hat{r}_{m_i}$ in Eq. (3.12)) vanishes for a linear approximation of the velocity field. This is the case for the simplicial elements used in this work.

3.1.4 Variational Form

Multiplying Eq. (3.22) by arbitrary (continuous) test functions q (with dimensions of pressure) and integrating over the analysis domain Ω gives

$$\int_\Omega -\frac{q}{\kappa}\frac{\partial p}{\partial t}d\Omega - \int_\Omega q\frac{\rho}{\kappa}\frac{\partial^2 p}{\partial t^2}d\Omega + \int_\Omega qd^v d\Omega + \int_\Omega q\tau\frac{\partial\hat{r}_{m_i}}{\partial x_i}d\Omega = 0 \quad (3.23)$$

Integrating by parts the last integral in Eq. (3.23) (and neglecting the space changes of τ) yields

$$\int_\Omega -\frac{q}{\kappa}\frac{\partial p}{\partial t}d\Omega - \int_\Omega q\tau\frac{\rho}{\kappa}\frac{\partial^2 p}{\partial t^2}d\Omega + \int_\Omega qd^v d\Omega - \int_\Omega \tau\frac{\partial q}{\partial x_i}\hat{r}_{m_i}d\Omega + \underbrace{\int_\Gamma q\tau\hat{r}_{m_i}n_i d\Gamma}_{\text{BT}} = 0 \quad (3.24)$$

where n_i are the components of the unit normal vector to the external boundary Γ of Ω.

Using Eq. (3.11) an equivalent form for the boundary term BT of Eq. (3.24) is obtained as

$$BT = \int_\Gamma q\tau\hat{r}_{m_i}n_i d\Gamma = \int_\Gamma q\tau\left(\rho\frac{\partial v_i}{\partial t} + \frac{2\mu}{3}\frac{\partial d^v}{\partial x_i}\right)n_i d\Gamma = \int_\Gamma q\tau\left(\rho\frac{\partial v_n}{\partial t} + \frac{2\mu}{3}\frac{\partial d^v}{\partial n}\right)d\Gamma \quad (3.25)$$

where $\frac{\partial d^v}{\partial n}$ is the derivative of the volumetric strain rate in the direction of the normal to the external boundary and v_n is the velocity normal to the boundary.

The term $\frac{2\mu}{3}\frac{\partial d^v}{\partial n}$ can be approximated as follows

$$\frac{2\mu}{3}\frac{\partial d^v}{\partial n} = \frac{2}{h_n}\left(\frac{2}{3}\mu^+ d^{v+} - \frac{2}{3}\mu^- d^{v-}\right) \quad \text{at } \Gamma \quad (3.26)$$

where (μ^+, d^{v+}) and (μ^-, d^{v-}) are respectively the values of μ and d^v at exterior and interior points of the boundary Γ and h_n is a characteristic length in the normal direction to the boundary. Figure 3.1 shows an example of the computation of $\frac{2}{3}\mu\frac{\partial d^v}{\partial n}$ at the side of a 3-noded triangle adjacent to the external boundary. The same procedure applies for 4-noded tetrahedra.

Clearly, at external boundaries $d^{v+} = 0$ and $d^{v-} = d^v$. Hence, d^{v-} coincides with the volumetric strain in the 3-noded triangular element adjacent to the boundary.

Fig. 3.1 Computation of the term of $\frac{2}{3}\mu\frac{\partial d^v}{\partial n}$ at the side ij of a 3-noded triangle ijk adjacent to the external boundary Γ [9]

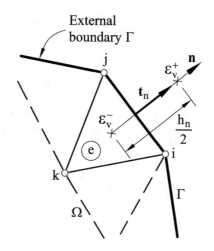

Using above argument Eq. (3.26) simplifies to

$$\frac{2\mu}{3}\frac{\partial d^v}{\partial n} = -\frac{4\mu}{3h_n}d^v \quad \text{at } \Gamma \tag{3.27}$$

On the other hand, the stresses at any boundary satisfy the traction equilibrium condition

$$\sigma_{ij}n_j - t_i = 0 \quad \text{at } \Gamma \tag{3.28}$$

Substituting Eqs. (3.3) and (3.4) into (3.28) and multiplying all terms by n_i, yields

$$2\mu\frac{\partial v_n}{\partial n} - \frac{2}{3}\mu d^v + p - t_n = 0 \quad \text{at } \Gamma \tag{3.29}$$

where t_n is the normal traction to the boundary (Fig. 3.1) and $\frac{\partial v_n}{\partial n} = n_i\frac{\partial v_i}{\partial x_j}n_j$.
From Eq. (3.29)

$$\frac{2}{3}\mu d^v = 2\mu\frac{\partial v_n}{\partial n} + p - t_n \quad \text{at } \Gamma \tag{3.30}$$

Substituting Eq. (3.30) into (3.27) and this one into Eq. (3.25), gives the expression of the boundary integral of Eq. (3.24) as

$$BT = \int_{\Gamma} q_\tau\left(\rho\frac{\partial v_n}{\partial t} - \frac{2}{h_n}\left(2\mu\frac{\partial v_n}{\partial n} + p - t_n\right)\right)d\Gamma \tag{3.31}$$

The normal velocity v_n is fixed at a Dirichlet boundary Γ_v and, hence, $\frac{\partial v_n}{\partial t} = 0$ at Γ_v. Also, accepting that $d^v = 0$ at Γ_v, the surface tractions at Γ_v coincide precisely with the reactions computed as $t_n = 2\mu\frac{\partial v_n}{\partial n} + p$. Hence, the boundary integral can be neglected at a Dirichlet boundary and, therefore, it has a meaning at a Neumann boundary Γ_t only. In conclusion,

$$BT = \int_{\Gamma_t} q\tau \left(\rho \frac{\partial v_n}{\partial t} - \frac{2}{h_n} \left(2\mu \frac{\partial v_n}{\partial n} + p - t_n \right) \right) d\Gamma \qquad (3.32)$$

Substituting Eq. (3.32) into (3.24) and using the expression of $\hat{r}_{m_i}$ of Eq. (3.12) yields the variational expression of the stabilized mass balance equation, after rearranging the different terms, as

$$\int_\Omega \frac{q}{\kappa} \frac{\partial p}{\partial t} d\Omega + \int_\Omega q\tau \frac{\rho}{\kappa} \frac{\partial^2 p}{\partial t^2} d\Omega - \int_\Omega q d^v d\Omega + \int_\Omega \tau \frac{\partial q}{\partial x_i} \left(2\mu \frac{\partial}{\partial x_i} (d_{ij}) + \frac{\partial p}{\partial x_i} + b_i \right) d\Omega$$
$$- \int_{\Gamma_t} q\tau \left[\rho \frac{\partial v_n}{\partial t} - \frac{2}{h_n} \left(2\mu \frac{\partial v_n}{\partial n} + p - t_n \right) \right] d\Gamma = 0$$

$$(3.33)$$

Expression (3.33) holds for 2D and 3D problems.

The terms involving the first and second material time derivative of the pressure and the boundary term in Eq. (3.33) are important to preserve the consistency of the residual form of the FIC mass balance equation. This, in turn, is essential for preserving the conservation of mass in the transient solution of free flow problems. The form of Eq. (3.33) is a key contribution of the new FIC-based stabilized formulation, versus previous works on this topic [7, 8, 24, 30, 31].

At an unloaded free surface (Neumann) boundary $t_n = 0$, and hence

$$BT = \int_{\Gamma_t} q\tau \left(\rho \frac{\partial v_n}{\partial t} - \frac{2}{h_n} \left(2\mu \frac{\partial v_n}{\partial n} + p \right) \right) d\Gamma \qquad (3.34)$$

For an inviscid fluid $\mu = 0$ and Eq. (3.34) simplifies to

$$BT = \int_{\Gamma_t} q\tau \left(\rho \frac{\partial v_n}{\partial t} - \frac{2p}{h_n} \right) d\Gamma \qquad (3.35)$$

Accounting for the term $\frac{\partial v_n}{\partial t}$ in the boundary integral of Eqs. (3.33)–(3.35) has proven to be relevant for the enhanced conservation of mass in free surface flows [9]. On the other hand, the effect of the term involving $\frac{\partial^2 p}{\partial t^2}$ was negligible in all the problems solved in this work [9].

Equation (3.33) is the starting point for deriving a new class of linear triangles with discontinuous pressure field adequate for analysis of incompressible flows with heterogeneous material properties [32].

The discretization of the mass stabilized equation is performed as it has been shown for the continuity equation in the previous chapter. So the analysis domain into finite elements is discretized using 3-noded linear triangles ($n = 3$) for 2D problems and 4-noded tetrahedra ($n = 4$) for 3D problems with local linear shape functions N_I defined for each node I ($I = 1, n$) of an element e.

3.1.5 FEM Discretization and Matrix Form

Substituting the approximations (2.9) and (2.62) into Eq. (3.33) and choosing a Galerkin form with $q = N_I$ gives the discretized form of the stabilized mass balance equation, after eliminating the arbitrary test functions as

$$\int_\Omega \frac{1}{\kappa} N^T N \frac{D\bar{p}}{Dt} d\Omega + \int_\Omega \frac{\tau\rho}{\kappa} N^T N \frac{D^2\bar{p}}{Dt^2} d\Omega - \int_\Omega N^T m^T B \bar{v} d\Omega +$$
$$+ \int_\Omega \tau(\nabla N)^T \nabla N \bar{p} d\Omega + \int_{\Gamma_t} \frac{2\tau}{h^n} N^T N \bar{p} d\Gamma - f_p = 0 \quad (3.36)$$

where

$$B_I = \begin{bmatrix} \frac{\partial N_I}{\partial x} & 0 & 0 \\ 0 & \frac{\partial N_I}{\partial y} & 0 \\ 0 & 0 & \frac{\partial N_I}{\partial z} \\ \frac{\partial N_I}{\partial y} & \frac{\partial N_I}{\partial x} & 0 \\ \frac{\partial N_I}{\partial z} & 0 & \frac{\partial N_I}{\partial x} \\ 0 & \frac{\partial N_I}{\partial z} & \frac{\partial N_I}{\partial y} \end{bmatrix}, \nabla N^T \equiv \begin{bmatrix} \nabla N_1 \\ \nabla N_2 \\ \cdots \\ \nabla N_N \end{bmatrix} \text{ with } \nabla = \begin{Bmatrix} \frac{\partial}{\partial x_1} \\ \frac{\partial}{\partial x_2} \\ \frac{\partial}{\partial x_3} \end{Bmatrix} \quad (3.37)$$

$$\text{and } N = [N_1, N_2, \ldots, N_N]^T$$

Equation (3.36) can be written in matrix form as

$$M_1 \dot{p} + M_2 \ddot{p} - Q^T \bar{v} + (L + M_b)\bar{p} - f_p = 0 \quad (3.38)$$

The matrices and vectors in Eq. (3.38) are assembled from the element contributions as

$$M_{1_{IJ}} = \int_\Omega \frac{1}{\kappa} N_I N_J d\Omega \quad (3.39)$$

$$M_{2_{IJ}} = \int_{\Omega} \tau \frac{\rho}{\kappa} N_I N_J d\Omega \tag{3.40}$$

$$M_{b_{IJ}} = \int_{\Gamma_t} \frac{2\tau}{h_n} N_I N_J d\Gamma \tag{3.41}$$

$$L_{IJ} = \int_{\Omega} \tau (\nabla^T N_I) \nabla N_J d\Omega \tag{3.42}$$

$$Q_{IJ} = \int_{\Omega} B_I^T m N_J d\Omega \tag{3.43}$$

$$f_{p_I} = \int_{\Gamma_t} \tau N_I \left[\rho \frac{D v_n}{Dt} - \frac{2}{h_n} (2\mu d_n - t_n) \right] d\Gamma - \int_{\Omega^e} \tau \nabla^T N_I b d\Omega \tag{3.44}$$

The boundary terms in vector f_p (Eq. (3.44)) can be incorporated with the matrices of Eq. (3.38). This, however, leads to a not symmetrical set of equations. For this reason these boundary terms are computed iteratively within the incremental solution scheme.

In a free surface fluid the presence in Eq. (3.38) of matrix M_b (Eq. (3.41)) enables the computation of the pressure without the need of prescribing its value at the free surface. This eliminates the error introduced when the pressure is prescribed to zero in free boundaries, which leads to considerable mass losses for viscous flows. Matrix M_b was introduced into the discretized stabilized mass balance equation in [16] using a fractional step method and heuristic arguments.

3.2 Solution Scheme for Quasi-incompressible Newtonian Fluids

The solution scheme for quasi-incompressible Newtonian fluids has a structure very similar to the two-step procedure presented for a general compressible material in Chap. 2. The linear momentum equations are solved for the increment of the velocities and the pressures are obtained from the continuity equation in the updated configuration. However, for dealing with the incompressibility, the stabilized form of the continuity equation (Eq. (3.38)) has to be considered. Furthermore, the linearized form of the momentum equations (Eq. (2.49)) has to be modified according to the constitutive law of Newtonian fluids.

3.2.1 Governing Equations

Linear Momentum Equations

For the sake of clarity, the general linearized form of the momentum equations is recalled. For each iteration i the following linear system is solved

$$K^i \Delta \bar{v} = R^i (^{n+1}\bar{v}^i, {}^{n+1}\sigma'^i, {}^{n+1}p^i) \tag{3.45}$$

where:

$$K^i = K^m (^{n+1}\bar{x}^i, c^{\sigma,i}) + K^g (^{n+1}\bar{x}^i, \sigma^i) + K^p (^{n+1}\bar{x}^i) \tag{3.46}$$

with

$$K^m_{IJ} = \int_\Omega B^T_I \Delta t \left[c^\sigma \right] B_J d\Omega \tag{3.47}$$

$$K^g_{IJ} = I \int_\Omega \beta^T_I \Delta t \sigma \beta_J d\Omega \tag{3.48}$$

$$K^p_{IJ} = I \int_\Omega N_I \frac{2\rho}{\Delta t} N_J d\Omega d\Omega \tag{3.49}$$

and

$$
{}^{n+1}R_{Ii} = \int_\Omega N_I \rho N_J d\Omega \, {}^{n+1}\bar{v}_{Ji} + \int_\Omega \frac{\partial N_I}{\partial x_j} {}^{n+1}\sigma'_{ij} d\Omega +
$$
$$
+ \int_\Omega \frac{\partial N_I}{\partial x_j} \delta_{ij} N_J d\Omega \, {}^{n+1}\bar{p}_J - \int_\Omega N_I {}^{n+1}b_i d\Omega - \int_{\Gamma_t} N_I {}^{n+1}t^p_i d\Gamma \tag{3.50}
$$

In order to obtain the linearized form of the momentum equations for quasi-incompressible Newtonian fluids, the tangent constitutive tensor in matrix K^m (3.47) and the Cauchy stress tensor that appears in the residual vector R and in K^g have to be computed using the adequate constitutive law.

For a Newtonian fluid, the stresses can be computed as

$$\sigma = \sigma' + pI = 2\mu d' + pI \tag{3.51}$$

For quasi-incompressible materials Eq. (3.2) holds within a general time interval $[n, n+1]$. For clarity purposes, the relation is rewritten as

$$^{n+1}p = {}^n p + \Delta t \kappa^{n+1} d^v \tag{3.52}$$

Considering a time interval $[n, n+1]$ and substituting Eq. (3.52) into (3.51) yields

$$^{n+1}\boldsymbol{\sigma} = \left(2\mu\mathbf{I}' + \Delta t\kappa \mathbf{I} \otimes \mathbf{I}\right) : \boldsymbol{d} +\,^{n}p\mathbf{I} \tag{3.53}$$

where the fourth-order tensor $\mathbf{I}'$ has been defined in Eq. (2.82).

For convenience, Eq. (3.53) is rewritten in the following form

$$^{n+1}\Delta\boldsymbol{\sigma} =\,^{n+1}\boldsymbol{\sigma} -\,^{n}\boldsymbol{\sigma} = \left(\boldsymbol{c}^{d} + \boldsymbol{c}^{\kappa}\right) : \boldsymbol{d} \tag{3.54}$$

where the following substitutions have been done:

$$^{n}\boldsymbol{\sigma} =\,^{n}p\mathbf{I} \tag{3.55}$$

$$\boldsymbol{c}^{d} = 2\mu\mathbf{I}' \tag{3.56}$$

$$\boldsymbol{c}^{\kappa} = \Delta t\kappa \mathbf{I} \otimes \mathbf{I} = \Delta t\kappa \boldsymbol{m}\boldsymbol{m}^{T} \tag{3.57}$$

The goal is to obtain a relationship between the measure of rate of stress and the rate of deformation in the form of Eq. (2.35). Equation (3.54) shows that, according to the constitutive law for Newtonian fluids, the rate of deformation is related to the Cauchy stress and not to rate of the Cauchy stress, as for hypoelastic solids (Eq. (2.83)). For this reason, in fluids preserving the objectivity of the stress rate measures is not a critical issue as for hypoelastic solids. Rigid rotations do not cause any state of stress, because the Cauchy stress is directly obtained from the rate of deformation. For these reasons, the rate of Cauchy stress can be simply defined from Eq. (3.54) as

$$\boldsymbol{\sigma}^{\nabla} = \dot{\boldsymbol{\sigma}} = \frac{\Delta\boldsymbol{\sigma}}{\Delta t} = \boldsymbol{c}^{\sigma} : \boldsymbol{d} = \left(\frac{\boldsymbol{c}^{d}}{\Delta t} + \frac{\boldsymbol{c}^{\kappa}}{\Delta t}\right) : \boldsymbol{d} \tag{3.58}$$

Note that Eq. (3.58) has the same structure as Eq. (2.35). For convenience, the Newtonian tangent moduli $\boldsymbol{c}^{\sigma}$ for the rate of stress is computed as the sum of the deviatoric and volumetric parts. Hence, substituting $\boldsymbol{c}^{\sigma}$ into Eq. (3.47) yields that for a quasi-incompressible Newtonian fluid the material part of the tangent matrix can be written as

$$\boldsymbol{K}^{m}_{Nf} = \boldsymbol{K}^{\mu} + \boldsymbol{K}^{\kappa} \tag{3.59}$$

where $\boldsymbol{K}^{\mu}$ and $\boldsymbol{K}^{\kappa}$ are defined for a generic finite element e and the pair of nodes I, J as

$$\boldsymbol{K}^{\mu}_{IJ} = \int_{\Omega^{e}} \boldsymbol{B}^{T}_{I} \left[\boldsymbol{c}^{\mu}\right] \boldsymbol{B}_{J} d\Omega \tag{3.60}$$

$$\boldsymbol{K}^{\kappa}_{IJ} = \int_{\Omega^{e}} \boldsymbol{B}^{T}_{I} \boldsymbol{m} \Delta t\kappa \boldsymbol{m}^{T} \boldsymbol{B}_{J} d\Omega \tag{3.61}$$

$$\text{with} \quad c^{\mu} = 2\mu \begin{bmatrix} 2/3 & -1/3 & -1/3 & 0 & 0 & 0 \\ & 2/3 & -1/3 & 0 & 0 & 0 \\ & & 2/3 & 0 & 0 & 0 \\ & & & 1/2 & 0 & 0 \\ \text{Sym.} & & & & 1/2 & 0 \\ & & & & & 1/2 \end{bmatrix}$$

The volumetric part of the tangent matrix K^{κ} can compromise the conditioning of the linear system because its terms are orders of magnitude larger than the viscous part. In order to prevent the numerical instabilities originated by the ill-conditioning of the tangent matrix, a reduced pseudo-bulk modulus, computed from the actual bulk modulus κ as $\kappa_p = \theta\kappa$, can be used in the expression of K^{κ} without altering the numerical results [33]. An adequate selection of the pseudo-bulk modulus also improves the overall accuracy of the numerical solution and the preservation of mass for large time steps [33]. For fully incompressible fluids ($\kappa = \infty$), a finite value of κ is used in K^{κ} as this helps to obtaining an accurate solution for velocities and pressure with reduced mass loss in few iterations per time step [33]. These considerations, however, do not affect the value of κ within matrix M_1 in Eq. (3.38). Clearly, the value of the terms of K^{κ} can also be limited by reducing the time step size. This, however, increases the overall computational cost. Another approach for improving mass conservation in incompressible flows was proposed in [31]. In Sect. 3.4.3 the details about this procedure will be given.

FIC Stabilized Mass Balance Equation

The pressure is obtained from the stabilized form of the mass balance equation (Eq. (3.38)). Introducing the time integration of the pressure (Eqs. (2.64) and (2.65)) into Eq. (3.38), yields

$$H \bar{p}^{i+1} = F_p(\bar{v}, \bar{p}) \tag{3.62}$$

where

$$H = \left(\frac{1}{\Delta t} M_1 + \frac{1}{\Delta t^2} M_2 + L + M_b \right) \tag{3.63}$$

and

$$F_p = \frac{M_1}{\Delta t} {}^n\bar{p} + \frac{M_2}{\Delta t^2} \left({}^n\bar{p} + {}^n\dot{\bar{p}}\Delta t \right) + Q^T \bar{v} + f_p \tag{3.64}$$

where all the matrices and the vectors of Eqs. (3.62)–(3.64) have been defined in Eqs. (3.39)–(3.44)

3.2.2 Solution Scheme

The complete solution scheme for a quasi-incompressible Newtonian fluid is
described for a generic time interval $[n, n + 1]$ in Box 8.

For each iteration i:

1. Compute the nodal velocity increments $\Delta \bar{v}$:

$$K^i \Delta \bar{v} = R^i (^{n+1}\bar{v}^i, \,^{n+1}\bar{p}^i)$$

where:

$$K^i = K^\mu(^{n+1}\bar{x}^i, c^\mu) + K^\kappa(^{n+1}\bar{x}^i, c^\kappa) + K^g(^{n+1}\bar{x}^i, \sigma^i) + K^\rho(^{n+1}\bar{x}^i)$$

2. Update the nodal velocities: $^{n+1}\bar{v}^{i+1} = {}^{n+1}\bar{v}^i + \Delta \bar{v}$

3. Update the nodal coordinates: $^{n+1}\bar{x}^{i+1} = {}^{n+1}\bar{x}^i + \bar{u}(\Delta \bar{v})$

4. Compute the nodal pressures $\bar{p}^{i+1}$:

$$H^{n+1} \bar{p}^{i+1} = F_p(^{n+1}\bar{v}^{i+1}, \,^{n+1}\bar{p}^i)$$

where: $H = \left(\dfrac{1}{\Delta t} M_1 + \dfrac{1}{\Delta t^2} M_2 + L + M_b \right)$ and

$$F_p = \frac{M_1}{\Delta t}{}^n\bar{p} + \frac{M_2}{\Delta t^2}\left({}^n\bar{p} + {}^n\bar{p}\Delta t \right) + Q^{T \, n+1}\bar{v}^{i+1} + f_{p,i}$$

5. Compute the updated stress measures:

$$^{n+1}\sigma'^{i+1} = 2\mu d'(^{n+1}\bar{v}^{i+1}) \quad ; \quad ^{n+1}\sigma^{i+1} = {}^{n+1}\sigma'^{i+1} + {}^{n+1}p^{i+1} I$$

6. Check convergence: $\| ^{n+1}R^{i+1}(^{n+1}\bar{v}^{i+1}, \,^{n+1}\bar{p}^{i+1}) \| < tolerance$

If condition 6 is not fulfilled, return to 1 with $i \leftarrow i + 1$.

Box 8. Iterative solution scheme for quasi-incompressible Newtonian fluids

In Box 9 all matrices and vectors that appear in Box 8 are given.

Vectors and matrices for the linear momentum equations

$$K^p_{IJ} = I \int_\Omega N_I \frac{2\rho}{\Delta t} N_J d\Omega \,, \; K^g_{IJ} = I \int_\Omega \beta^T_I \Delta t \sigma \beta_J d\Omega$$

$$K^\mu_{IJ} = \int_\Omega B^T_I \left[c^\mu \right] B_J d\Omega \,, \; K^\kappa_{IJ} = \int_\Omega B^T_I m \kappa \Delta t m^T B_J d\Omega$$

$$c^\mu = \mu \begin{bmatrix} 4/3 & -2/3 & -2/3 & 0 & 0 & 0 \\ & 4/3 & -2/3 & 0 & 0 & 0 \\ & & 4/3 & 0 & 0 & 0 \\ & & & 1 & 0 & 0 \\ \text{Sym.} & & & & 1 & 0 \\ & & & & & 1 \end{bmatrix}$$

Vectors and matrices for the continuity equation

$$M_{1_{IJ}} = \int_\Omega \frac{1}{\kappa} N_I N_J d\Omega \,, \; M_{2_{IJ}} = \int_\Omega \tau \frac{\rho}{\kappa} N_I N_J d\Omega \,, \; M_{b_{IJ}} = \int_{\Gamma_t} \frac{2\tau}{h_n} N_I N_J d\Gamma$$

$$L_{IJ} = \int_\Omega \tau (\nabla N_I) \nabla N_j d\Omega \,, \; Q_{IJ} = \int_\Omega B^T_I m N_J d\Omega$$

$$f_{p_I} = \int_{\Gamma_t} \tau N_I \left[\rho \frac{D v_n}{Dt} - \frac{2}{h_n} (2\mu d_n - t_n) \right] d\Gamma - \int_\Omega \tau \nabla^T N_I b d\Omega$$

with $\tau = \left(\dfrac{8\mu}{h^2} + \dfrac{2\rho}{\delta} \right)^{-1}$

where κ is the bulk modulus and μ is the viscosity.

Box 9. Element form of the matrices and vectors of Box 8

3.3   Solution Scheme for Quasi-incompressible Hypoelastic Solids

The differences between the solution scheme for quasi-incompressible and compressible hypoelastic solids concern only the continuity equation. In fact, the same linearized form of the continuum equations derived in the previous chapter for the compressible hypoelastic model within the mixed Velocity–Pressure formulation holds also for the quasi-incompressible scheme. However, for dealing with incompressibility the scheme needs to be stabilized.

In this work the FIC stabilization procedure originally derived for Newtonian fluids in [9] and proposed again in Sect. 3.1 of this chapter, has been tested for the first time for quasi-incompressible hypoelastic solids. In order to use the same form of Eqs. (3.62)–(3.64) for hypoelastic quasi-incompressible solids, the fluid parameters (the fluid viscosity μ_f and the fluid bulk modulus κ_f) should be substituted with the equivalent solid parameters. The similarity between Newtonian fluids and hypoelastic solids is evident comparing the computation of the Cauchy stress tensor increment for the two mentioned constitutive laws.

For quasi-incompressible Newtonian fluids Eq. (3.54) holds and, for clarity purposes, is here rewritten as

$$^{n+1}\Delta\boldsymbol{\sigma}_f = 2\mu_f\mathbf{I}' : \boldsymbol{d} + \Delta t\kappa_f\boldsymbol{I} \otimes \boldsymbol{I} : \boldsymbol{d} \tag{3.65}$$

From Eqs. (2.96), the increment of the Cauchy stress for hypoelastic solids is

$$^{n+1}\Delta\boldsymbol{\sigma}_s = 2\Delta t\mu_s\mathbf{I}' : \boldsymbol{d} + \Delta t\kappa_s\boldsymbol{I} \otimes \boldsymbol{I} : \boldsymbol{d} \tag{3.66}$$

where bulk modulus for the solid κ_s is computed from the Lamè constants μ_s and λ_s or in terms of the Young modulus and the Poisson ratio of the material (Eqs. (2.77), (2.78) and (2.81)).

Equations (3.65) and (3.66) show the duality between hypoelastic and Newtonian quasi-incompressible constitutive laws. In the latter the deviatoric and the volumetric parts of the Cauchy stress tensor are controlled by the dynamic viscosity μ_f and the bulk modulus κ_f, respectively. The equivalent roles in hypoelastic solids are taken by the second Lamè constant scaled by the time increment $(\Delta t\mu_s)$ and the bulk modulus κ_s. Thanks to this equivalence, the stabilized mass continuity equation derived for quasi-incompressible fluids (Eq. (3.38)) holds also for quasi-incompressible hypoelastic solids just by replacing the fluid parameters μ_f and κ_f with the equivalent terms for hypoelastic solids $\Delta t\mu_s$ and κ_s.

Note that using this analogy the differences between the incremental solution schemes for Newtonian fluids and quasi-incompressible hypoelastic solids are minimal and they essentially differ on the way to compute the stresses and the tangent moduli according to the specific constitutive law. For hypoelastic quasi-incompressible model, the stress tensor is computed with Eq. (2.99) and the tangent moduli is obtained from Eq. (2.86). The solid finite element implemented with this formulation is called VPS-element.

In Box 10, the iterative solution incremental scheme for quasi-incompressible hypoelastic solids using the stabilized mixed Velocity–Pressure formulation is described for a generic time increment $[^n t, ^{n+1} t]$.

For each iteration i:

1. Compute the nodal velocity increments $\Delta\bar{v}$:

$$K^i \Delta\bar{v} = R^i(^{n+1}\bar{v}^i, {}^{n+1}\sigma'^i, {}^{n+1}\bar{p}^i)$$

where $K^i = K^m(^{n+1}\bar{x}^i, c^{\sigma J,i}) + K^g(^{n+1}\bar{x}^i, \sigma^i) + K^p(^{n+1}\bar{x}^i)$

2. Update the nodal velocities: $^{n+1}\bar{v}^{i+1} = {}^{n+1}\bar{v}^i + \Delta\bar{v}$

3. Update the nodal coordinates: $^{n+1}\bar{x}^{i+1} = {}^{n+1}\bar{x}^i + \bar{u}(\Delta\bar{v})$

4. Compute the nodal pressures $\bar{p}^{i+1}$:

$$H_s^{n+1}\bar{p}^{i+1} = F_p(^{n+1}\bar{v}^{i+1}, {}^{n+1}\bar{p}^i)$$

where: $H_s = \left(\dfrac{1}{\Delta t}M_{1s} + \dfrac{1}{\Delta t^2}M_{2s} + L_s + M_{bs}\right)$ and

$$F_p = \dfrac{M_{1s}}{\Delta t}{}^n\bar{p} + \dfrac{M_{2s}}{\Delta t^2}\left(^n\bar{p} + {}^n\dot{\bar{p}}\Delta t\right) + Q^{T\ n+1}\bar{v}^{i+1} + f_{ps,i}$$

5. Compute the updated stress measures:

$$^{n+1}\sigma^{i+1} = {}^n\hat{\sigma} + {}^{n+1}\Delta\bar{p}^{i+1}I + 2\mu\Delta t\left[I' : d\left(\bar{v}^{i+1}\right)\right]$$

6. Check convergence: $\|\ ^{n+1}R^{i+1}(^{n+1}\bar{v}^{i+1}, {}^{n+1}\bar{p}^{i+1})\ \| < tolerance$

If condition 6 is not fulfilled, return to 1 with $i \leftarrow i + 1$.

Box 10. Iterative solution scheme for quasi-incompressible hypoelastic solids

In Box 11 all matrices and vectors that appear in Box 10 are given.

Vectors and matrices for the linear momentum equations

$$K_{IJ}^p = I \int_\Omega N_I \frac{2\rho}{\Delta t} N_J d\Omega \ , \ K_{IJ}^g = I \int_\Omega \beta_I^T \Delta t \sigma \beta_J d\Omega$$

$$K_{IJ}^m = \int_\Omega B_I^T \Delta t \left[c^{\sigma J} \right] B_J d\Omega$$

$$c^{\sigma J} = \begin{bmatrix} \kappa_s + \frac{4}{3}\mu_s & \kappa_s - \frac{2}{3}\mu_s & \kappa_s - \frac{2}{3}\mu_s & 0 & 0 & 0 \\ \kappa_s - \frac{2}{3}\mu_s & \kappa_s + \frac{4}{3}\mu_s & \kappa_s - \frac{2}{3}\mu_s & 0 & 0 & 0 \\ \kappa_s - \frac{2}{3}\mu_s & \kappa_s - \frac{2}{3}\mu_s & \kappa_s + \frac{4}{3}\mu_s & 0 & 0 & 0 \\ 0 & 0 & 0 & \mu_s & 0 & 0 \\ 0 & 0 & 0 & 0 & \mu_s & 0 \\ 0 & 0 & 0 & 0 & 0 & \mu_s \end{bmatrix}$$

Vectors and matrices for the continuity equation

$$M_{1s_{IJ}} = \int_\Omega \frac{1}{\kappa_s} N_I N_J d\Omega \ , \ M_{2s_{IJ}} = \int_\Omega \tau \frac{\rho}{\kappa_s} N_I N_J d\Omega \ , \ M_{bs_{IJ}} = \int_{\Gamma_t} \frac{2\tau}{h_n} N_I N_J d\Gamma$$

$$L_{s_{IJ}} = \int_\Omega \tau (\nabla N_I) \nabla N_J d\Omega \ , \ Q_{IJ} = \int_\Omega B_I^T m N_J d\Omega$$

$$f_{ps_I} = \int_{\Gamma_t} \tau N_I \left[\rho \frac{Dv_n}{Dt} - \frac{2}{h_n}(2\Delta t \mu_s d_n - t_n) \right] d\Gamma - \int_\Omega \tau \nabla^T N_I b d\Omega$$

with $\tau_s = \left(\dfrac{8\Delta t \mu_s}{h^2} + \dfrac{2\rho}{\delta} \right)^{-1}$

where $\kappa_s = \frac{2\mu_s}{3} + \lambda_s$ and λ_s and μ_s are the Lamè constants

Box 11. Element form of the matrices and vectors of Box 10

3.4 Free Surface Flow Analysis

Free surface flows are those fluids with at least one side in contact with the air. Their numerical solution is critical because the free surface contours change continuously and they need to be tracked in order to solve accurately the boundary value problem. In this work, this task is carried out using the Lagrangian finite element procedure called *Particle Finite Element Method (PFEM)* [34]. In the PFEM the mesh nodes are treated as the fluid particles. As a consequence, the free surface contours are automatically detected by the nodes positions. The PFEM will be described and analyzed in detail in Sect. 3.4.1.

Another typical drawback associated to free surface flows is the deterioration of the mass preservation of the numerical method. The indetermination of the free surface position introduces in the scheme an additional mass loss source to those induced by the inaccuracy of the numerical method. In Sect. 3.4.2 the mass conservation feature of the PFEM-FIC stabilized formulation is studied in detail and many validation examples are given.

In Sect. 3.4.3 the key role of the pseudo bulk modulus κ_p for guaranteeing the mass preservation of the quasi-incompressible formulation will be highlighted. In fact, the actual value of the bulk modulus κ deteriorates the conditioning of the linear system. This may affect the mass preservation feature and even the convergence of the whole scheme. Using a reduced value for the bulk modulus, computed as $\kappa_p = \theta\kappa$, all these inconveniences are overcome. In the same section a practical strategy for computing the pseudo bulk modulus is also given and tested with several numerical examples.

3.4.1 The Particle Finite Element Method

The *Particle Finite Element Method* (*PFEM*) is a Lagrangian finite element procedure ideated and developed by Idelsohn and Oñate and their work group [10, 34–36]. The PFEM addresses those problems where severe changes of topology occur. This may happen in free surface fluid dynamics, in non-linear solid mechanics, in FSI problems or in thermal coupled problems. The essential idea of the PFEM is to follow the topology of the deformed bodies by moving the nodes of the mesh according to the equations of motion in a Lagrangian way. In other words, in the PFEM the mesh nodes are treated as particles and they transport their momentum together with all their physical properties. All this produces the deformation of the finite elements discretization that needs to be rebuilt whenever a threshold value for the distortion is reached. Remeshing is one of the most characteristic points of the PFEM [37]. This operation is performed via an efficient combination of the Delaunay Tessellation [38, 39] and the Alpha Shape method [11]. Once the mesh is generated, the differential problem is integrated again over the new mesh in the classical FEM fashion.

In the following parts of this section the essential points of the PFEM will be highlighted. Specifically, first the remeshing procedure will be described, then the basic steps of the PFEM will be summarized and finally the advantages and disadvantages of the technique will be highlighted.

3.4.1.1 Remeshing

In order to solve accurately the FEM problem, at each time step it has to be guaranteed a good quality for the mesh. In fact, even a single degenerated element may compromise the entire computation. For this reason the remeshing algorithm must be reliable and robust. In the PFEM this is guaranteed through an efficient algorithm

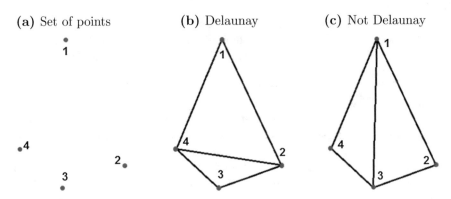

Fig. 3.2 2D Delaunay triangulation of a set of points

that combines two trustworthy techniques, the Delaunay triangulation and the Alpha Shape method.

Given a cloud of points in the space, the Delaunay tessellation guarantees the creation of the most homogenous discretization for those points. This is obtained by ensuring that the circumball, circumcircle in 2D, built on all the nodes of a simplex does not contain any other node of some other simplex [40]. Actually this is guaranteed only in 2D where it is proved that there exists a unique Delaunay triangulation. In 3D there may be the eventuality that the criterion is not fulfilled for certain tetrahedra. In 2D the Delaunay triangulation has the property of maximizing the minimum inner angle of all the triangles of the mesh (this is not guaranteed in 3D). Note that the maximization of the minimum angle does not imply the minimization of the maximum angle, see Fig. 3.2.

This strategy may be helpful for avoiding the creation of degenerated elements as the *slivers* (simplices with null surface or volume) but it is not sufficient for their complete elimination. It will be explained later that further controls and modifications on the cloud of points are required for this objective.

The Delaunay triangulation alone cannot ensure the detection of the physical boundaries of the bodies. For this purpose, the so-called Alpha Shape method is applied on the tessellation obtained with the Delaunay procedure. The role of the Alpha Shape method is to erase those simplices that are excessively distorted or big. In order to do this, the mean mesh size h and the circumradius of each element r_e are computed. Next the following check is done for all the elements

$$if \; r_e > \alpha h \quad \rightarrow \quad erase \; the \; e - element \qquad (3.67)$$

where α is a parameter that is chosen arbitrarily.

The selection of α is arbitrary. However, it must be taken into account the following considerations. If α is excessively large, the Alpha Shape method is worthless because no element will be erased. This means that the Delaunay triangulation remains unchanged. On the other hand, an excessive small value for α induces the

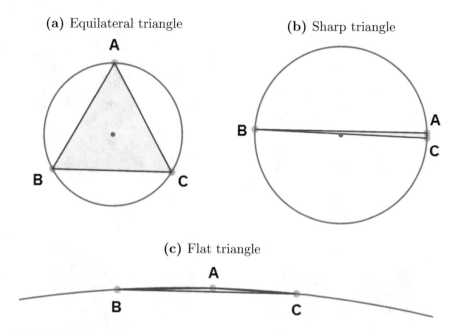

(a) Equilateral triangle

A

B

C

(b) Sharp triangle

B

A

C

(c) Flat triangle

A

B

C

Fig. 3.3 Triangles with the same mean size and their circumcircles

elimination of all the simplices. Specifically the lowest admissible α parameter is the one of the equilateral triangle for which $\alpha = 1/\sqrt{3}$. In Fig. 3.3 the circumcircle is plotted for simplices with the same mean length for edges. Given a mean size for the edges of a simplex, the equilateral triangle is the one with the minimum circumradius. Equally, for a given circumradius, the equilateral triangle is the simplex which edges have the maximum mean length.

The arbitrariness in the choice of the α parameter may affect the numerical results. In fact, two different values for α may give two different configurations. See for example the case of Fig. 3.4. For the same Delaunay triangulation (Fig. 3.4b), the configuration obtained with α_1 (Fig. 3.4c) is different from the one obtained with α_2 (Fig. 3.4d), where $\alpha_1 > \alpha_2$. Specifically, the mesh obtained with α_1 accepts a larger number of elements from the Delaunay discretization than the one given by α_2, especially in the free surface zone. However, in the section dedicated to the validation examples it will be shown that the method is convergent. This means that the error introduced by a certain α reduces with the refinement of the mesh.

The remeshing can be performed also in the opposite order, hence using first an Alpha Shape method for detecting the contours of the domains and then performing the Delaunay triangulation with the geometric restrictions of those boundaries [36, 41, 42]. This can be done by performing the Delaunay triangulation via the *advancing front* technique [43].

Both remeshing algorithms may generate particles or elements separated from the rest of the domain. Remember that all the information is stored in the nodes, so the

(a) Set of points   **(b)** Delaunay triangulation

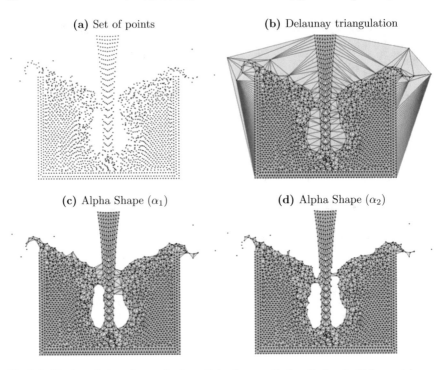

(c) Alpha Shape (α_1)   **(d)** Alpha Shape (α_2)

Fig. 3.4 2D triangulation of a set of points. Alpha shape method applied on the Delaunay triangulation ($\alpha_1 > \alpha_2$)

governing equations can be computed also for an isolated particle. This allows, for example, the simulation of detached fluid drops.

However, the first strategy is preferable because it leads to a reduced computational cost and it requires a lower implementation effort. For these reasons, in this work the remeshing is performed applying the Alpha Shape technique after the Delaunay triangulation.

Neither the Alpha Shape method nor the Delaunay triangulation can prevent from the creation of highly distorted elements. Let consider, for example, the triangle depicted in Fig. 3.3b. The circumradius associated to this type of triangle is not excessively large and it may fulfill Eq. (3.67). The Alpha Shape method and the Delaunay triangulation can only ensure the best physical mesh (so the best discretization that respects the physical boundaries of the domain) for a given cloud of points. However, during the motion, the nodes distribution may be such that it is impossible to avoid the presence of poor quality elements. In order to avoid this eventuality the remeshing with the Delaunay tessellation and the Alpha Shape method must be combined with some additional controls.

For example, for avoiding the presence of sharp elements in the triangulation, as the one shown in Fig. 3.5, one should monitor the edge lengths of each triangle.

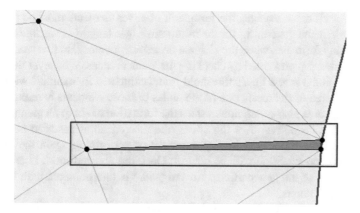

Fig. 3.5 Distorted element in a 2D triangulation

Fig. 3.6 Sliver element in
3D

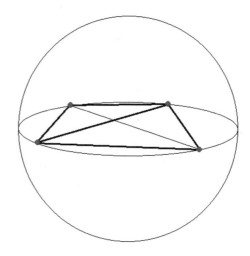

Whenever the distance between two nodes is smaller than a prearranged critical length, one of those nodes is removed and placed in other zone of the mesh, for example where there are some distorted elements. The reallocation of the nodes is done with the purpose of preserving the number of particles and, consequently, guaranteeing the conservations of the mean mesh size (if the volume does not change).

Another type of critical elements are the slivers, so the simplices with null area or volume. These may form when there are three points contained in the same line, or four points laying in the same 3D plane as shown in Fig. 3.6.

In theory, the Alpha Shape method should not allow the formation of those elements because the associated circumradius is huge, see Fig. 3.3c. However, these elements are critical from the numerical point of view and they can escape to the Alpha Shape control. Furthermore, also in the case that the sliver element is erased from the Delaunay triangulation, this would create a not physical void within the domain. For all these reasons, the formation of sliver elements must be prevented. For these elements the control for the minimum edges lengths is worthless because the sliver may form also when the element nodes are not close. In this work, the formation of slivers is prevented by checking the element areas. Whenever the element surface or volume is less than a threshold value, computed for example with respect to the mean area of the mesh, one of the nodes of those elements is removed. Note that this control is also beneficious versus the formation of sharp elements.

After these operations over the nodes, the discretization needs to be updated. This can be done modifying locally the mesh or performing again the Delaunay triangulation and the Alpha Shape control. Once the updated mesh is created, the nodal variables of the new particles are assigned, via interpolation, with the values of the neighbor nodes.

All these criteria have been explained for homogeneous meshes. However, they can be easily extended to heterogeneous meshes. For example, if there is a mesh refinement in a specific zone of the domain, that refined mesh can be preserved if the described checks for the minimum nodes distance and the elements surface are done not for the whole mesh but for the local mesh.

3.4.1.2   Basic Steps

The basic steps of the PFEM algorithm are here given with the help of a graphic representation (Fig. 3.7). Consider a domain V containing fluid and solid subdomains.

In this work the PFEM is used only for the fluid domains. The solids are computed with the classical FEM and they maintain the same discretization for the whole

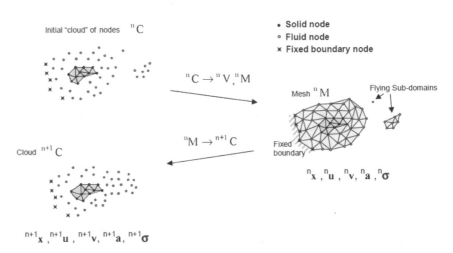

Fig. 3.7 Sequence of steps to update a "cloud" of fluid nodes and a discretized solid domain from time n ($t = {}^{n}t$) to time $n + 1$ ($t = {}^{n}t + \Delta t$)

duration of the analysis. In many other works the PFEM has been also used for modeling the solid and the fluid domains, indifferently [44–47].

At the beginning of a time step (first picture of Fig. 3.7) the fluid is represented by a cloud of points, or particles, that store all the information about the physics (for example the density and the viscosity of the fluid), the geometry (information about the local mesh size), the kinematics (the velocity and the acceleration) and the pressure. On the other hand, the solid has the same mesh of the previous time step and the history for its kinematics and the stress and strain fields is stored in both the nodes and the elements.

The first step represents the distinguishing step of the PFEM and it consists on the creation of a mesh for the fluid. The algorithm is the one that has been described in Sect. 3.4.1.1. So first the Delaunay triangulation is performed and then the physical contours of the domains are recognized with the Alpha Shape method. Specifically, the fluid detects automatically all its boundaries, hence the rigid walls, its free surface contours and the interface with the solid. During this step flying subdomains may form or/and some particles may detach from the rest of the domain (second picture of Fig. 3.7)

Once all the domains are discretized, it is possible to solve the equations of continuum mechanics for each of the subdomains using the FEM. The state variables are computed at the next (updated) configuration for $^n t + \Delta t$: velocities, pressure and viscous stresses in the fluid and displacements, stresses and strains in the solid. After that, the fluid mesh can be erased and these steps are repeated. For reducing the computational cost associated to the detection of the fluid boundaries it is very useful to memorize those nodes that in the previous time step were located at the contours. In this way, the nodes checked by the Alpha Shape method can be drastically reduced.

Once again, note that in the fluid all the information is stored in the nodes so this operation does not require interpolation procedures for recovering the elemental information. If the PFEM would be used also for the solid mechanics analysis, an algorithm for transferring the elemental information from the previous to the new mesh should be implemented. In [48] an efficient technique for this purpose is described and validated.

3.4.1.3 Advantages and Disadvantages

The key differences between the PFEM and the classical FEM are the remeshing technique and the identification of the domain boundary at each time step. The quality of the numerical solution depends on the discretization chosen as in the standard FEM and, as it has been explained, adaptive mesh refinement techniques can be used to improve the solution. So from this point of view, during the computation step, the PFEM behaves as a classical Lagrangian FEM, with the same advantages and drawbacks.

However with the PFEM the remeshing may introduce a further source of error in the numerical scheme. In fact during this operation some new elements may be created and some other erased. These changes cause local perturbations of the equilibrium reached at the previous time step and they may affect the accuracy and the convergence of the scheme. In fact at the beginning of the time step, each node stores the kinematics and the pressure obtained with a discretization that may be different than the new one. So those values that at the end of the previous step were at equilibrium for the previous configuration and mesh, may not be at equilibrium at the beginning of the new time step for the new configuration created by the remeshing.

Another drawback induced by the remeshing concerns the variation of volume. Although, generally, the volumes gained and lost with the remeshing are compensated, there may be cases where the variation of volume is not negligible. From the personal experience of the author, remeshing tends to increase the volume of the domain. For avoiding this inconvenience, some additional criteria for limiting the creation of new elements should be added. Generally, the critical zones are the ones near to the free surface, where, typically, the remeshing tends to create new elements that 'close' the surface waves and increase the overall volume of the fluid, as shown in Fig. 3.8.

It is possible to reduce this tendency of the remeshing by locally penalizing the Alpha Shape method. In this work, the free surface nodes have assigned a smaller Alpha Shape parameter α than the other nodes of the mesh.

This penalization is also good for improving the timing of the contact with the rigid walls or the solid interfaces. In particular, this strategy may delay the phenomenon illustrated in Fig. 3.9. The elements created at the free surface may accelerate the impact between the fluid streams and the containing walls. The smaller is α for the interface nodes, the later a contact element between those nodes and the solid or rigid

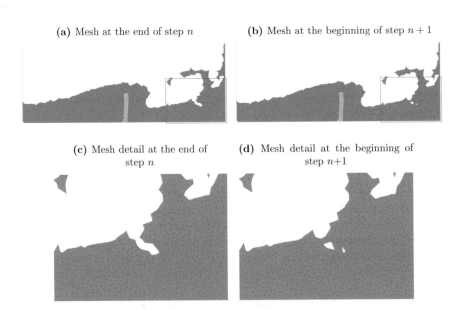

(a) Mesh at the end of step n

(b) Mesh at the beginning of step $n + 1$

(c) Mesh detail at the end of step n

(d) Mesh detail at the beginning of step $n+1$

Fig. 3.8 Example of variation of volume induced by the remeshing

(a) End of time step n **(b)** Beginning of time step $n+1$

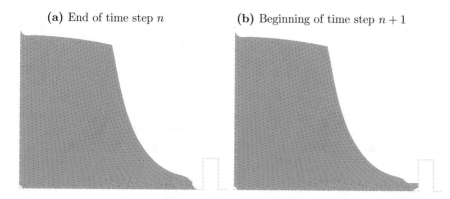

Fig. 3.9 Example of detection of contact through the remeshing

contour is built. In fact, a smaller α for the interface nodes delays the formation of the contact elements with the rigid (or deformable) contours reproducing better the physical phenomenon of the contact. However, the Alpha Shape cannot be penalized excessively because otherwise the fluid could pass through the wall, or the solid interface.

The remeshing also represents an additional computational cost. In previous works it has been found that the computational time associated to the remeshing grows linearly with the number of nodes [25]. Specifically, for a single processor Pentium IV PC the meshing consumes for 3D problems around 15 % of the total CPU time per time step, while the solution of the equations (with typically 3 iterations per time step) and the system assembly consumes approximately 70 and 15 % of the CPU time per time step, respectively. The treatment of the boundary nodes is an issue that deserves a particular attention. In the previous works with the PFEM stick conditions were generally used for the nodes on the rigid walls. However, this may induce pressure concentrations on the boundary elements and deteriorate locally the quality of the mesh. In fact the stick condition affects a zone that has the same order of magnitude of the spatial discretization. This may induce a non-physical behavior, especially when inviscid fluids are analyzed or/and coarse meshes are employed. The problem is even more critical in 3D. Recently, Cremonesi et al. [49, 50] used slip conditions in the simulation of landslides for better modeling the interaction between a landslide and the substrate interface. In the mentioned works, the wall nodes are treated in an Eulerian way by adding the convective term for the boundary elements. In this work, the wall particles are still computed in a Lagrangian framework and they are free to move along the direction of the walls. The slip conditions are simulated with a simple algorithm. Essentially it consists on leaving the wall particles move along the direction of the wall until when the separation from the original position is larger than a prearranged critical distance. In that case, during the remeshing procedure, the particle is removed and reallocated at its original position. The reallocation of the particle is done in order to prevent the creation of voids in the walls that may cause fluid leakage. The kinematics and the pressure of the moved particle are obtained via

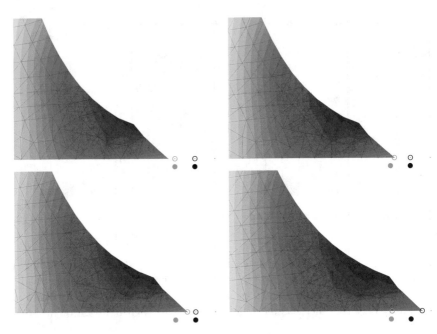

Fig. 3.10 Algorithm for moving the wall particles

interpolation from the neighbor nodes, in the classical PFEM fashion. In Fig. 3.10 an example of the application of the algorithm is given. The pictures refer to the front of a flow.

The void circles follow the position of two nodes of the wall discretization, the full circles denote the original position of those nodes. The pictures show that the wall nodes are free to move along the wall direction and when they reach the maximum separation from the original position (in this case $0.6h_w$, where h_w is the distance between the node of the wall) they are reallocated at the original position. In the section dedicated to the validation examples, some interesting comparisons between stick and slip problems are given, both in 2D and 3D. It will be shown that this algorithm can be very helpful for preserving the quality of the mesh and for the overall accuracy of the computation.

All the weak points of the PFEM presented until now are the price to pay in order to gain all the benefits that this Lagrangian technique can give. The PFEM in fact makes simple some tasks that are extremely critical for a complex computational analysis and, in some cases, it allows the solution and the modeling of problems that other methods cannot even face.

The PFEM is designed for those problems where a huge change of topology occurs. The mesh nodes define the evolution of a domain during its history of deformation. In this way, the detection of the free surface in a fluid flow or the contours of a melting solid object do not introduce in the scheme any additional complication or implementation work because their contours are known at the beginning of each time step.

With the PFEM the bodies are totally free to deform and move in the space. There is not any limitation in this sense because the domain is continuously updated. Hence the final configuration can be very different then the initial one and occupy a volume of space arbitrary bigger than the initial one. This cannot been done with a classical Eulerian strategy, where a bounding box needs to be defined at the beginning of the analysis.

Furthermore with the PFEM the interface between a fluid stream and a solid is detected automatically via the same Alpha Shape method that generates the mesh. The interface is formed by the nodes of those hybrid elements (elements which nodes belong to both the solid and the fluid) that have been generated by the Delaunay triangulation and they fulfill the Alpha Shape criteria.

PFEM is also easy to couple with other numerical methods. This thesis is an example of the coupling of the PFEM with the FEM. In other works, it has been shown that the PFEM can also be easily coupled with discrete methods [42]. Furthermore, within the same PFEM, one may chose the preferred numerical method for solving the governing equations after the remeshing. In this work a mixed Velocity–Pressure stabilized formulation is used for the fluid, but one may implement any other Lagrangian methodology.

All this explains why in the past the PFEM has been employed for facing challenging simulations as tunneling [44, 45], forming processes [26, 51], melting of polymers [46, 52], transport erosion and sedimentation in fluids [25], and fluid-multibody interaction [27].

In conclusion, the PFEM is an extremely powerful technology which shortcomings are legitimized by the complexity of the problems to solve.

3.4.2   Mass Conservation Analysis

This section is focused on the analysis of the mass preservation feature of the PFEM-FIC stabilized formulation for free surface flow problems.

It is useful to distinguish two different sources of variation of volume (in all the examples presented in this work, the density is a constant, so the concepts of mass variation and volume variation are totally equivalent). The first source of volume variation is induced by the numerical solution for each time step increment. The volume may vary during the non-linear iterations due to the inaccuracy of the scheme. So the volume of the fluid domain changes from a volume ^nV computed at the beginning of the time step to a slightly different volume $^{n+1}\bar{V}$ computed after the convergence of the solution scheme. This volume variation caused by the computation is called ΔV^c.

The second source of mass loss is related to the remeshing. After the achievement of the convergence the fluid domain is remeshed. During this process the volume may vary because some elements are erased or new elements are added to the mesh. The

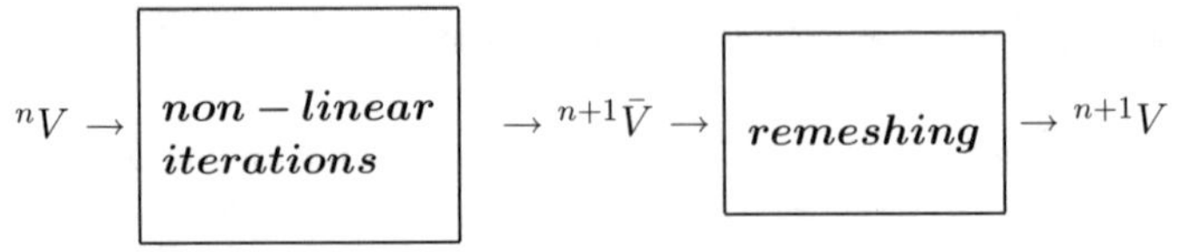

Fig. 3.11 Monitoring of volume within a time step interval $({}^n t, {}^{n+1} t)$

volume at the end of the time step is ${}^{n+1} V$ and the variation induced by the remeshing with respect to ${}^{n+1} \bar{V}$ is named ΔV^m. In Fig. 3.11 a graphical representation of the scheme is given.

The sum of both contributions gives the total volume variation for each time step such as

$$\Delta V^{tot} = \Delta V^c + \Delta V^m \tag{3.68}$$

where

$$\Delta V^c = {}^{n+1} \bar{V} - {}^n V \tag{3.69}$$

$$\Delta V^m = {}^{n+1} V - {}^{n+1} \bar{V} \tag{3.70}$$

For the proper understanding of the mass preservation feature of a formulation is essential to distinguish these different mass losses sources. Nevertheless, it is improper to define these two sources as uncorrelated. In fact, the variation of mass generated by the remeshing may affect the mass losses of the non-linear scheme. In fact, the remeshing induces a perturbation of the equilibrium reached in the previous time step through the elimination and creation of elements. This may affect the convergence of the numerical scheme and, consequently, the satisfaction of the incompressibility constraint.

The numerical examples presented in this section are taken from [9]. All the examples have been run using the real value of the bulk modulus κ in matrix $\boldsymbol{K}^\kappa$ (Eq. (3.61)). This matrix is the component of the tangent matrix $\boldsymbol{K}$ for linear momentum equations that takes into account the pressure variation. In Sect. 3.4.3 it will be shown that substituting in matrix $\boldsymbol{K}^\kappa$ (Eq. (3.61)) the real bulk modulus κ with a pseudo bulk modulus defined as $\kappa_p = \theta \kappa$, is helpful for the conditioning and the overall accuracy of the analysis.

3.4.2.1   Numerical Examples

In order to verify the mass preservation feature of the formulation, several problems for which guaranteeing the mass conservation is critical, have been solved. In this section, some of these involving impact and mixing of free surface fluids, are presented.

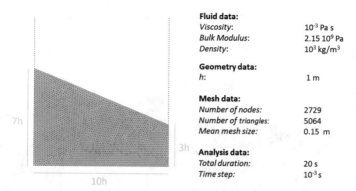

Fluid data:
Viscosity:	10^{-3} Pa s
Bulk Modulus:	2.15 10^9 Pa
Density:	10^3 kg/m^3

Geometry data:
h:	1 m

Mesh data:
Number of nodes:	2729
Number of triangles:	5064
Mean mesh size:	0.15 m

Analysis data:
Total duration:	20 s
Time step:	10^{-3} s

Fig. 3.12 2D analysis of sloshing of water in a prismatic tank. Initial geometry, analysis data and mesh of 5064 3-noded triangles discretizing the water in the tank [9]

Sloshing of Water in a Prismatic Tank

The problem has been solved first in 2D. Figure 3.12 shows the geometry of the tank, the material properties, the time step size and the initial mesh of 5064 3-noded triangles discretizing the interior fluid. The fluid oscillates due to the hydrostatic forces induced by its original position. In this and in the next problems, the effect of the surrounding air has not been taken into account.

This problem, as for all the ones presented in this section, has been solved using the real bulk modulus in matrix $\boldsymbol{K}^\kappa$ (Eq. (3.61)), that is equivalent to compute the pseudo bulk modulus $\kappa_p = \theta\kappa$ with the parameter $\theta = 1$.

Figures 3.13 and 3.14 show some snapshots of the water stream at different times. Pressure contours are plotted over the deformed configuration.

(a) $t = 5.7s$ (b) $t = 7.4s$

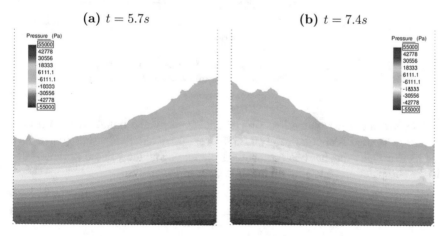

Fig. 3.13 2D sloshing of water in a prismatic tank. Snapshots of water geometry at two different times ($\theta = 1$). Colours indicate pressure contours [9] (I/II)

(a) $t = 13.3s$ **(b)** $t = 18.6s$

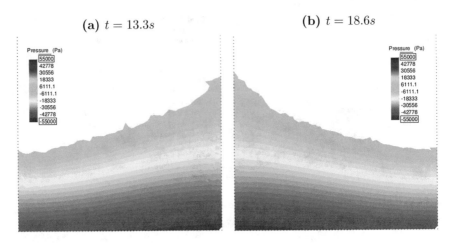

Fig. 3.14 2D sloshing of water in a prismatic tank. Snapshots of water geometry at two different times ($\theta = 1$). Colours indicate pressure contours [9] (II/II)

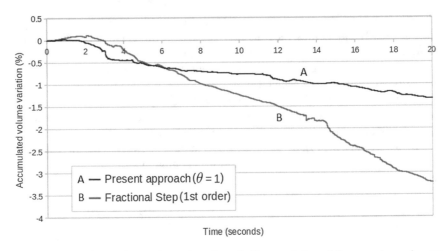

Fig. 3.15 2D sloshing of water in a prismatic tank. Time evolution of the percentage of water volume loss due to the numerical algorithm. Comparison with the fractional step solution [9]

The accumulated volume loss (in percentage versus the initial volume) for the method proposed is approximately 1.33 % over 20 s of simulation time (curve *A* of Fig. 3.15). This value has been computed by summing all the volume variations due to the computation (ΔV^c) for all the time steps of the analysis as

$$^n\Delta V_\% = \frac{\sum_{i=1}^{n} {}^i\Delta V^c}{V_{initial}} \cdot 100 \tag{3.71}$$

where $V_{initial}$ is the initial volume.

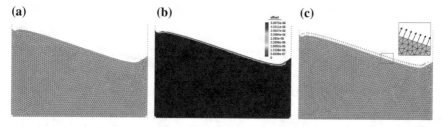

Fig. 3.16 Strategy for recovering the initial fluid volume by correcting the free surface at mesh generation level. **a** Compute total volume variation (ΔV) before remeshing. **b** Compute free surface offset $= \frac{\Delta V}{L_{\text{free surface}}}$. **c** Move free surface nodes in the normal direction to the boundary a distance equal to the offset computed in **b** [9]

In Fig. 3.15 also the mass losses obtained using a standard first order fractional step method and the PFEM [9] are plotted. Clearly the method proposed in this work leads to a reduced overall fluid volume loss.

In some cases, it may be required to ensure the absolute absence of volume variations. With a small shifting of the free surface it is possible to guarantee the perfect mass conservation in the fluid domain. The method consists on moving the free surface nodes in the normal direction with an offset equal to $\frac{\Delta V}{L_{\text{free surface}}}$, where the variation of volume ΔV is computed as the sum of the volume variation due to the remeshing at the previous time step $^{n-1}\Delta V^m$ and the volume variation due to computation of the current time step $^n\Delta V^c$. $L_{\text{free surface}}$ is the length of the free surface. In Fig. 3.16 a graphical representation of the technique in 2D is given.

This procedure induces an arbitrary alteration of the equilibrium configuration, thus it has to be used cautiously. A fundamental requirement for using this technique is that the solution scheme produces very small volume variations and so the correction is minimal.

Figure 3.17 shows that the total fluid volume loss can be reduced to almost zero by applying this strategy at each time step.

Figure 3.18 shows a comparison of the fluid volume losses between using the real bulk modulus ($\theta = 1$) and an arbitrarily reduced pseudo bulk modulus ($\theta = 0.08$). For both cases, the time step increment is $\Delta t = 10^{-3}$ s.

Figure 3.19 shows that a similar improvement in the volume preservation can be obtained using $\theta = 1$ and reducing the time step to $\Delta t = 10^{-4}$ s. This, however, increases the cost of the computations.

These results show that accurate numerical results with reduced volume losses can be obtained by appropriately adjusting the parameter θ in the tangent bulk modulus matrix, while keeping the time step size to competitive values in terms of CPU cost. In Sect. 3.4.3 this issue will be analyzed in detail. Specifically, the effects of the bulk modulus κ in terms of volume preservation and overall accuracy will be analyzed and an efficient strategy to predict the best value for the pseudo bulk modulus ($\kappa_p = \theta\kappa$) will be presented.

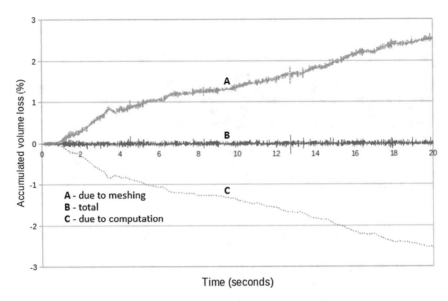

Fig. 3.17 2D sloshing of water in a prismatic tank ($\theta = 1$). Mass losses applying the strategy described in Fig. 3.16 [9]

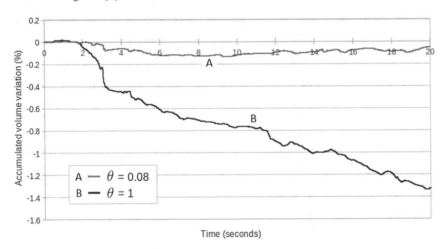

Fig. 3.18 2D sloshing of water in a prismatic tank. Time evolution of percentage of water volume loss obtained using the current method with $\theta = 0.08$ (*curve A*) and $\theta = 1$ (*curve B*) $\Delta t = 10^{-3}$ s [9]

In Fig. 3.20 the input data for the 3D case are given. The same sloshing problem of the 2D simulation is solved using a relative coarse initial mesh of 106771 4-noded tetrahedra and $\theta = 1$.

Figure 3.21 shows the results for the 3D analysis at the same time instants of Fig. 3.13.

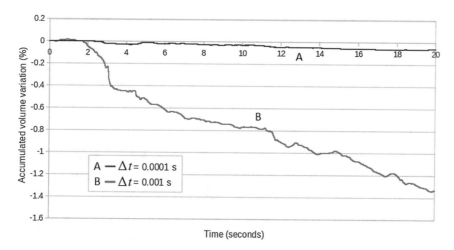

Fig. 3.19 2D sloshing of water in a prismatic tank. Time evolution of percentage of mass loss obtained with the current method. *Curve A $\theta = 1$ and $\Delta t = 10^{-4}$ s. Curve B $\theta = 1$ and $\Delta t = 10^{-3}$ s* [9]

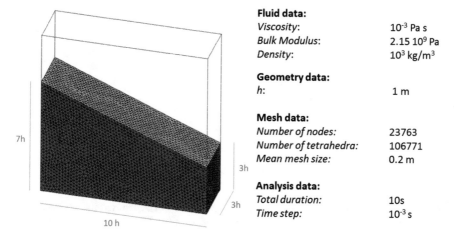

Fluid data:

Viscosity:	10^{-3} Pa s
Bulk Modulus:	$2.15\ 10^9$ Pa
Density:	10^3 kg/m^3

Geometry data:

h:	1 m

Mesh data:

Number of nodes:	23763
Number of tetrahedra:	106771
Mean mesh size:	0.2 m

Analysis data:

Total duration:	10s
Time step:	10^{-3} s

Fig. 3.20 3D analysis of sloshing of water in prismatic tank ($\theta = 1$). Initial geometry and analysis data [9]

In Figs. 3.22 and 3.23 the mass losses are analyzed. The values of Fig. 3.22 have been computed as Eq. (3.71). It is remarkable that the percentage of total fluid volume loss due to the numerical scheme after 10 s of analysis is approximately 1 %.

The values plotted in the graph of Fig. 3.23 are the percentage mass loss due to computation for each time step.

(a) $t = 5.7s$ (b) $t = 7.4s$

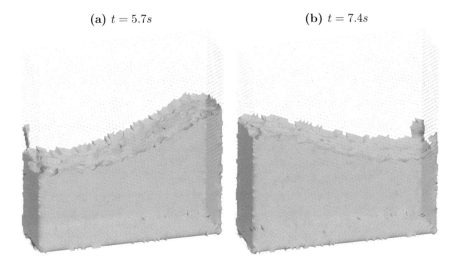

Fig. 3.21 3D analysis of sloshing of water in prismatic tank ($\theta = 1$). Snapshots of water geometry at two different times ($t = 5.7$ s *left* and $t = 7.4$ s *right*) [9]

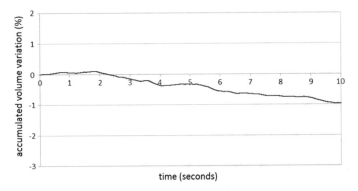

Fig. 3.22 3D analysis of sloshing of water in prismatic tank ($\theta = 1$). Time evolution of accumulated water volume loss (in percentage) due to the numerical algorithm [9]

These values have been computed for each time step n as

$$^{n}\Delta V_{\%} = \frac{^{n}\Delta V^{c}}{V_{initial}} \cdot 100 \qquad (3.72)$$

Collapse of Two Water Columns in a Prismatic Tank

This problem simulates the 2D motion, impact and subsequent mixing of two fluid streams originated by the collapse of two water columns located at the end sides of a prismatic tank. Figure 3.24 shows the initial geometry and the problem data.

The two water columns have been discretized with 3988 3-noded triangles.

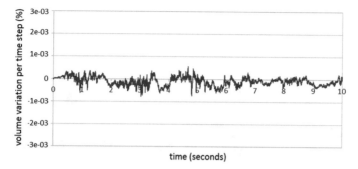

Fig. 3.23 3D analysis of sloshing of water in prismatic tank ($\theta = 1$). Volume loss per time step over 2 s of analysis. Average volume loss per time step: 1.64×10^{-4} % [9]

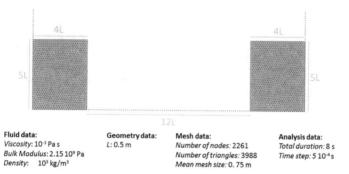

Fluid data:
Viscosity: 10^{-3} Pa s
Bulk Modulus: $2.15 \, 10^9$ Pa
Density: 10^3 kg/m³

Geometry data:
L: 0.5 m

Mesh data:
Number of nodes: 2261
Number of triangles: 3988
Mean mesh size: 0.75 m

Analysis data:
Total duration: 8 s
Time step: $5 \, 10^{-4}$ s

Fig. 3.24 Collapse and impact of two water columns in a prismatic tank. Analysis data, initial geometry and discretization (3988 3-noded triangles) [9]

The effect of the surrounding air has not been taken into account in the analysis. The problem has been solved for $\theta = 1$.

Figure 3.25 shows four snapshots of the motion of the water columns after removal of the retaining walls. Alter a few instants the two water streams impact with each other and mix as shown in the figures.

The evolution of the percentage of the initial fluid volume loss over the simulation time is shown in Fig. 3.26.

The values plotted in Fig. 3.26 have been computed using Eq. (3.71). A maximum of 2.8 % of the initial fluid volume is lost over eight seconds of analysis. This can be considered a low value for a problem of this complexity.

In Fig. 3.27 the mass losses for each time step computed with Eq. (3.72) are plotted.

Falling of a Water Sphere in a Cylindrical Tank Containing Water

The final example is the 3D analysis of the impact and mixing of a water drop as it falls in a cylindrical tank filled with the same fluid. Figure 3.28 shows the material and analysis data and the initial discretization (88892 4-noded tetrahedra).

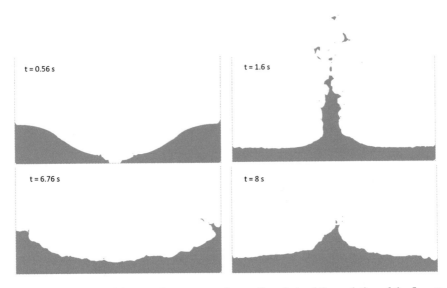

Fig. 3.25 Collapse and impact of two water columns. Snapshots of the evolution of the flow at different times [9]

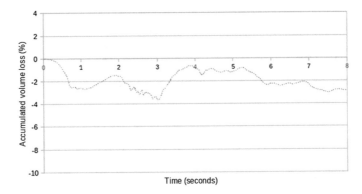

Fig. 3.26 Collapse and impact of two water columns. Accumulated fluid volume (in percentage) over eight seconds of analysis due to the numerical algorithm. Results for $\theta = 1$ [9]

The problem has been solved using the real bulk modulus also in the tangent matrix (i.e. taking $\theta = 1$). Figure 3.29 shows four snapshots of the mixing process at different times.

The total water mass lost in the sphere and the tank due to the numerical algorithm was $\simeq 2\,\%$ after 3 s of analysis (Fig. 3.30). The average volume loss per time step was $2.54 \times 10^{-4}\,\%$.

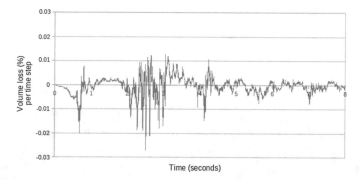

Fig. 3.27 Collapse and impact of two water columns. Volume loss (in %) per time step. Results for $\theta = 1$ [9]

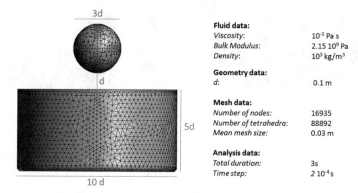

Fluid data:
Viscosity:	10^{-3} Pa s
Bulk Modulus:	$2.15\ 10^9$ Pa
Density:	10^3 kg/m³

Geometry data:
d:	0.1 m

Mesh data:
Number of nodes:	16935
Number of tetrahedra:	88892
Mean mesh size:	0.03 m

Analysis data:
Total duration:	3s
Time step:	$2\ 10^{-4}$ s

Fig. 3.28 Falling of a water sphere in a tank filled with water. Analysis data, initial geometry and discretization (88892 4-noded tetrahedra) [9]

3.4.3 Analysis of the Conditioning of the Solution Scheme

This part is devoted to study the effect of the bulk modulus in the iterative matrix (matrix $\boldsymbol{K}$ (Eq. (3.46)) on the partitioned solution scheme for quasi-incompressible fluids. For clarity purposes, the decomposition of matrix $\boldsymbol{K}$ is recalled. $\boldsymbol{K}$ is computed as

$$\boldsymbol{K} = \boldsymbol{K}^\rho + \boldsymbol{K}^g + \boldsymbol{K}^\mu + \boldsymbol{K}^\kappa \tag{3.73}$$

with

$$\boldsymbol{K}^g_{IJ} = \boldsymbol{I} \int_\Omega \beta_I^T \Delta t \sigma \beta_J d\Omega \tag{3.74}$$

$$\boldsymbol{K}^\rho_{IJ} = \boldsymbol{I} \int_\Omega N_I \frac{2\rho}{\Delta t} N_J d\Omega \tag{3.75}$$

Fig. 3.29 Falling of a water sphere in tank containing water. Evolution of the impact and mixing of the two liquids at different times. Results for $\theta = 1$ [9]

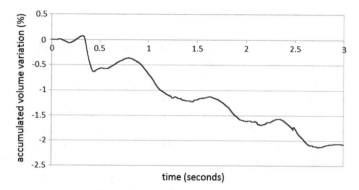

Fig. 3.30 Falling of a water sphere in a tank containing water ($\theta = 1$). Accumulated volume over three seconds of analysis due to the numerical algorithm [9]

$$K^{\mu}_{IJ} = \int_{\Omega^e} B^T_I c^{\mu} B_J d\Omega \tag{3.76}$$

$$K^{\kappa}_{IJ} = \int_{\Omega^e} B^T_I m \Delta t \kappa m^T B_J d\Omega \tag{3.77}$$

In this section, it is shown that matrix K^{κ} (Eq. (3.77)) can deteriorate the conditioning of the linear system (Eq. (3.45)) and the overall accuracy of the numerical scheme if the real bulk modulus κ is used. In order to avoid these drawbacks, a pseudo bulk modulus κ_p should be used. In this section, a practical rule to set up the value of a pseudo-bulk modulus a priori in matrix K^{κ} for improving the conditioning of the linear system is presented. The efficiency of the proposed strategy is tested in several problems analyzing the advantage of the modified bulk tangent matrix for the stability of the pressure field, the convergence rate and the computational speed of the analyses. The technique has been tested on the FIC/PFEM Lagrangian formulation presented, but it can be easily extended to other quasi-incompressible stabilized finite element formulations. The method proposed is based on using a pseudo-bulk modulus κ_p in the volumetric component K^{κ} (Eq. (3.77)) of the linear momentum tangent matrix K (Eq. (3.73)), while the actual physical value of the bulk modulus κ is used for the numerical solution of the mass conservation equation, namely in matrices M_1 (Eq. (3.39)) and M_2 (Eq. (3.40)). The pseudo-bulk modulus κ_p is defined "a priori" as a proportion of the actual bulk modulus of the fluid (i.e. $\kappa_p = \theta \kappa$ with $0 < \theta \leq 1$). The study here presented recalls the ideas presented in [33].

This section is structured as follows. In the first part, the numerical inconveniences induced by the real bulk modulus are analyzed. Then the strategy for predicting the optimum value for pseudo bulk modulus is explained. Finally several numerical examples are given in order to validate the technique.

3.4.3.1 Drawbacks Associated to the Real Bulk Modulus

The linearized system (Eq. (3.45)) suffers from numerical instabilities due to the ill-conditioning of the iteration matrix K (Eq. (3.73)) caused by the presence of the bulk modulus in matrix K^κ (Eq. (3.77)), so in the part of K that takes into account the pressure variation. This problem was already pointed out in previous works where similar partitioned schemes were used [31, 53, 54]. The ill-conditioning of the iterative matrix of the linear momentum equations originates from the different orders of magnitude of its two main contributions, namely, the mass matrix K^ρ (Eq. (3.75)) and the bulk matrix K^κ (for low viscous flows the contributions of matrices K^μ (Eq. (3.76)) and K^g (Eq. 3.74)) are negligible). Typically the terms of the bulk matrix are orders of magnitude larger than those of the mass matrix.

A reliable measure of the quality of a matrix is the *condition number* [55]. For a general matrix A, the condition number is defined as

$$C = cond(A) = \| A \| \cdot \| A^{-1} \| \tag{3.78}$$

where $\| A \|$ denotes here the L2 norm of matrix A.

The condition number C gives an indication of the results accuracy of the linear equation solution obtained by matrix inversion. Values of C close to 1 indicate a well-conditioned matrix.

The deterioration of the quality of matrix K affects directly the convergence of the iterative linear solver. In this work, the iterative *Bi-Conjugate Gradient* (BCG) solver has been used and its tolerance has been fixed to 10^{-6}.

In this solution scheme, the time step increment affects highly the conditioning of the iterative matrix K, as it has been shown in the first numerical example of Sect. 3.4.2.1. In fact, K^ρ is inversely proportional to the time increment, while K^κ depends linearly on it (see Eqs. (3.75) and (3.77)). For this reason, K is well-conditioned only for a tight range of time increments.

For the sake of clarity, a numerical example is used to visualize and quantify the inconveniences caused by the ill-conditioned matrix in the iterative solution scheme. The problem chosen is the 2D water sloshing in a prismatic tank presented in Sect. 3.4.2.1. However, with the purpose of reducing the computational cost of the analysis, the problem has been solved using a coarse mesh. So the initial geometry and the material data are the same as in Fig. 3.12, but the mesh is the one shown in Fig. 3.31. In Table 3.1 all problem data are summarized.

To highlight the key role of the time step on the conditioning of the iterative matrix, the problem has been solved for two different time increments without reducing the bulk modulus in matrix K^κ ($\theta = 1$). For a time step of $\Delta t = 10^{-3}$ s, the iterative matrix has a condition number $C = 41$ while, for $\Delta t = 10^{-2}$ s, the condition number is $C = 3009$. If a larger time step is used, the linear system cannot even be solved. This deterioration is reflected by the number of iterations of the linear BCG solver: for the former case the average number of iterations is around 20, while for the latter it is around 177. Clearly, this also leads to a significant increase of computational time for solving a time step.

Fig. 3.31 2D water
sloshing. Initial geometry
and finite element mesh [33]

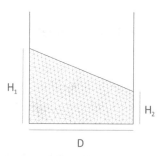

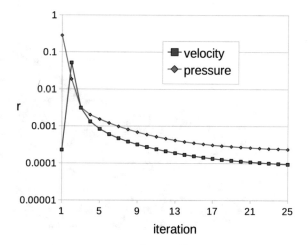

Table 3.1 2D water sloshing.
Problem data

Number of elements	705
Number of nodes	428
Average mesh size	0.4 m
H_1	7 m
H_2	3 m
D	10 m
Viscosity	10^{-3} Pa · s
Density	10^3 kg/m^3
Bulk modulus	$2.15 \cdot 10^9$ Pa

The ill-conditioning of the linear system also affects the rate of convergence of
the iterative loop of the scheme given in Box 8.

In the graph of Fig. 3.32 the convergence rates r for the pressure and the velocity
fields at $t = 1.75$ s are displayed.

Fig. 3.32 2D water sloshing
($\theta = 1$). Convergence of the
velocities and pressure at
$t = 1.75$ s [33]

For each iteration i of the non-linear loop, the convergence rates for the nodal velocities and pressures have been computed as

$$r_v = \frac{\|\Delta \bar{v}^{i+1}\|}{\|\bar{v}\|} \leq e_v \tag{3.79a}$$

$$r_p = \frac{\|\bar{p}^{i+1} - \bar{p}^i\|}{\|\bar{p}\|} \leq e_p \tag{3.79b}$$

where e_v and e_p are prescribed error norms for the nodal velocities and the nodal pressures, respectively. In the examples solved in this work, $e_v = e_p = 10^{-4}$ has been chosen.

The values of Fig. 3.32 have been obtained considering a time step increment of $\Delta t = 10^{-2}$ s. The curves show that the convergence of the scheme is slow, especially for the pressure field. In fact, the pressure error (Eq. (3.79b)) after 25 iterations is still larger than the pre-defined tolerance of $e_p = 10^{-4}$. The lack of a good convergence for the pressure field has two main negative effects: the pressure solution is not accurate and mass conservation is not preserved.

The pressure contours at $t = 1.75$ s are shown in Fig. 3.33.

Concerning the mass conservation of the fluid, Fig. 3.34 shows the accumulated mass variation in absolute value versus time. This values have been computed at each time step n as

$$^n\Delta V_\% = \frac{\sum_{i=1}^n |^i\Delta V^c|}{V_{initial}} \cdot 100 \tag{3.80}$$

where ΔV^c is the volume loss due to computation and $V_{initial}$ is the initial volume.

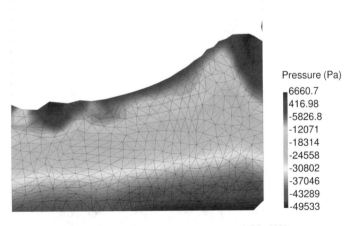

Pressure (Pa)

6660.7
416.98
-5826.8
-12071
-18314
-24558
-30802
-37046
-43289
-49533

Fig. 3.33 2D water sloshing ($\theta = 1$). Pressure contours at $t = 1.75$ s [33]

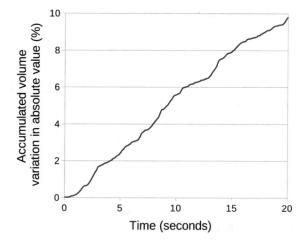

Fig. 3.34 2D water sloshing ($\theta = 1$). Accumulated percentage of mass variation in absolute value versus time [33]

After 20 s of simulation, the percentage of mass loss is 9.8 %, which corresponds to a mean volume variation of $4.9 \cdot 10^{-3}$ % per time step.

In conclusion, the use of real bulk modulus ($\theta = 1$) in matrix $\boldsymbol{K}^\kappa$ leads to ill-conditioned linear systems and, as consequence, it limits the range of suitable time step increments and deteriorates the convergence of the solution scheme.

These drawbacks were overcome in previous works [31, 54] by substituting the physical bulk modulus with a smaller pseudo bulk modulus κ_p. In this way, it is possible to extend the applicability of the partitioned scheme to a larger range of time increments. However, this strategy is based on heuristic criteria and cannot be used widely because each problem requires a specific value for κ_p. In other words, a pseudo bulk modulus that works well for certain analysis can fail for a different one. In this work, the optimum pseudo bulk modulus κ_p is computed a priori and the strategy for predicting its value will be explained in the next section.

3.4.3.2 Optimum Value for the Pseudo Bulk Modulus

In this work, the pseudo bulk modulus is computed as $\kappa_p = \theta\kappa$ choosing for the parameter θ a value able to guarantee the well-conditioning of the tangent matrix $\boldsymbol{K}$ (Eq. (3.77)). The numerical examples presented in this section will show that this reduced value for the volumetric compressibility improves the overall accuracy of the scheme and does not affect the mass conservation. The reason is that the modification only affects the quality of the iterative matrix and the rate of convergence of the linear momentum equations, while the continuity equation, through which the mass conservation constraint is imposed, is not modified. In other words, the physical value of the bulk modulus κ is used in matrices $\boldsymbol{M}_1$ (Eq. (3.39)) and $\boldsymbol{M}_2$ (Eq. (3.40)). This feature represents an innovation versus previous approaches where the pseudo bulk modulus is also used in the mass conservation equation [31, 53].

The parameter θ is computed as

$$\theta = \frac{mean(|\ \mathbf{K}^p\ |)}{mean(|\ \mathbf{K}^\kappa\ |)} \tag{3.81}$$

where the operator $(|\cdot|)$ denotes the mean of the absolute values of the non-zero matrix components.

For a uniform mesh of elements of characteristic size h, the here called 'optimum value' of θ is estimated as follows

$$\theta \approx \frac{2N_c^2 \cdot \rho \cdot h^2}{\kappa \cdot \Delta t^2} \tag{3.82}$$

where N_c is the value of the shape function N_I at the element center and the following approximation has been used:

$$\frac{\partial N_I}{\partial x} \approx \frac{1}{h} \tag{3.83}$$

If water is considered ($\rho = 10^3\ \mathrm{kg/m^3}$ and $\kappa = 2.15 \cdot 10^9\ \mathrm{Pa}$) and linear triangles are used ($N = 1/3$), θ has the following dependency with the mesh size and the time step,

$$\theta \approx 10^{-7} \cdot \left(\frac{h}{\Delta t}\right)^2 \tag{3.84}$$

Typically, the parameter θ is calculated at the beginning of the time step at the first iteration of the non-linear loop. It can also be computed at every time step, at certain instants of the analysis or only once at the beginning of the analysis. This is so because the order of magnitude of θ does not vary during the analysis, unless the time step is changed, or a refinement of the mesh is performed. In these cases, θ needs to be calculated again because it has a square dependency on both parameters h and Δt. The numerical results presented in this work have been obtained computing θ via Eq. (3.81) at the beginning of the analyses only.

θ can be computed and assigned locally to each element or globally to the whole mesh. If computed globally, θ is calculated only once for all the mesh and it is used for all the elemental contributions to the bulk iteration matrix. Otherwise, Eq. (3.81) is computed for each element of the mesh. The former approach has a reduced computational cost but it works worse than the local approach for not uniform finite element discretizations. In particular, it is recommended to use the local approach when a refinement of the mesh is performed (in the next section, a problem with a refined zone is studied). In this work, unless otherwise mentioned, the global approach for computing θ is used.

3.4.3.3   Numerical Examples

In order to show the efficiency of the method described in Sect. 3.4.3.2, two representative free surface problems involving the flow of water have been solved: the sloshing problem introduced in Sect. 3.4.3.1 and the collapse of a water column against a rigid obstacle presented in [56]. The so-called dam break problem has been chosen here to demonstrate that the strategy does not affect the incompressibility constraint at all and the method is able to simulate problems of impact of fluids. For both sloshing and dam break problems, a comparison of the performances of the scaled bulk matrix and the scheme with $\theta = 1$ is given.

The applicability and generality of the method is studied using very different average mesh sizes and time steps for both problems. Note that the dynamics of the sloshing problem is completely different from the dam break one.

Water Sloshing in a Tank

The problem of Fig. 3.31 is solved using the parameter θ computed as in Eq. (3.81). For $\Delta t = 10^{-3}$ s, $\theta = 0.154$, while for $\Delta t = 10^{-2}$ s, $\theta = 0.0053$. For both time step increments the resulting condition number of the iterative matrix is $C = 23$ and the average number of iterations of the linear BCG solver is around 15. The improvement versus using $\theta = 1$ is evident, if compared to the numbers presented in Sect. 3.4.3. Reducing the number of iterations of the linear solver, also reduces the computational time. For example, using $\Delta t = 10^{-2}$ s for a duration of the sloshing simulation of 20 s, the total computational time for $\theta = 1$ is 2746 s while for $\theta = 0.0053$ it reduces to 1600 s. Also the convergence of the non-linear loop improves with the optimum value of θ.

In Figs. 3.35 and 3.36 the convergence of the velocity and pressure fields, respectively, obtained with $\theta = 1$ and $\theta = 0.0053$, is compared. The faster convergence of the solution using a smaller value of θ is noticeable.

Fig. 3.35 2D water sloshing. Convergence of the velocities at $t = 1.75$ s for $\theta = 1$ and $\theta = 0.0053$ (optimum value) [33]

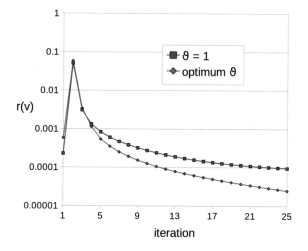

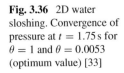

Fig. 3.36 2D water sloshing. Convergence of pressure at $t = 1.75$ s for $\theta = 1$ and $\theta = 0.0053$ (optimum value) [33]

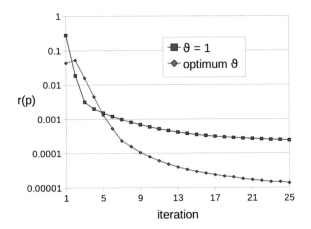

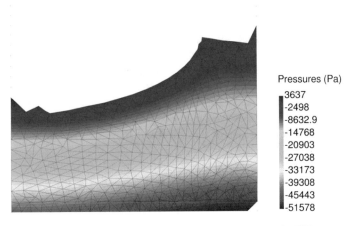

Fig. 3.37 2D water sloshing ($\theta = 0.0053$). Pressure contours at $t = 1.75$ s [33]

In Fig. 3.37 the pressure solution at time t $= 1.75$ s obtained with $\theta = 0.0053$ is illustrated. It can be appreciated a remarkable enhancement versus the solution for $\theta = 1$ (see Fig. 3.33). Note that the elements generated in the free surface region adjacent to the boundaries are due to the coarseness of the mesh and the remeshing criteria and not to the computation. A better result can be obtained using a smaller average mesh size or a refined mesh in the free surface region.

As stated in Sect. 3.4.3, the convergence in the continuity equation affects the conservation of mass of the fluid. It has been just shown that the convergence of the pressure improves using a good prediction of θ. Consequently, also the mass conservation is better ensured using $\theta = 0.0053$ than using $\theta = 1$. In the graph of Fig. 3.38 the percentage of accumulated mass variation (in absolute value) versus time obtained with $\theta = 1$ and $\theta = 0.0053$ are compared. The values plotted in the

Fig. 3.38 2D water sloshing. Accumulated mass variation in absolute value along the duration of the analysis. Solution for $\theta = 1$ and $\theta = 0.0053$ (optimum value) [33]

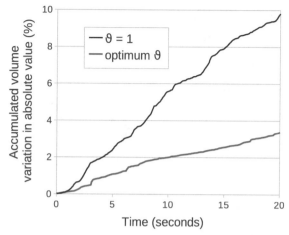

Fig. 3.39 2D water sloshing ($\theta = 0.0053$). Accumulated mass variation versus time [33]

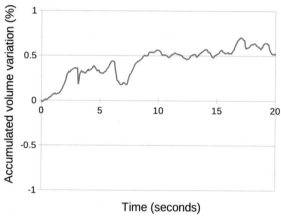

graph have been computed according to Eq. (3.80). The better mass preservation of the solution with the smaller value of θ is remarkable.

In the graph of Fig. 3.39 the accumulated mass variation obtained with $\theta = 0.0053$ is illustrated. The values plotted in the graph have been computed using Eq. (3.71). After 20 s of simulation, the scheme with the reduced value of θ has an accumulated mass variation of 0.52 %, which corresponds to a mean volume variation for each time step of $1.7 \cdot 10^{-3}$ %. The solution with $\theta = 0.0053$ guarantees a better conservation of mass than with $\theta = 1$. This represents another evidence that the (quasi)-incompressibility constraint is not affected by reducing the bulk modulus for solving the momentum equations.

Fig. 3.40 2D water sloshing. Number of iterations of the linear solver for different numbers of velocity degrees of freedom. Results for $\theta = 1$ and the optimum value of θ [33]

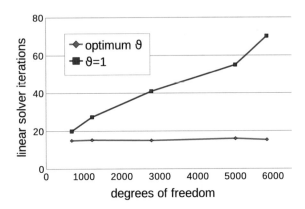

Table 3.2 2D water sloshing. Numerical values of the graph of Fig. 3.40 [33]

Average mesh size	Degrees of freedom (velocities)	Number of iterations	
		$\theta = 1$	Optimum θ
0.4	682	20	15 ($\theta = 0.535$)
0.3	1220	27	15 ($\theta = 0.304$)
0.2	2782	41	15 ($\theta = 0.136$)
0.15	5002	55	16 ($\theta = 0.0801$)
0.1	5840	70	15 ($\theta = 0.0361$)

Influence of the Mesh Size

In order to verify the applicability of the method, the sloshing problem has been solved using $\Delta t = 10^{-3}$ s and different mesh sizes. In particular, the following average mesh sizes have been used: $h = 0.1, 0.15, 0.2, 0.3$ and 0.4 m.

The problem was solved setting $\theta = 1$ and computing a priori its reduced value using Eq. (3.81).

The curves in Fig. 3.40 show that the reduced value of θ guarantees better results versus using $\theta = 1$. For all the meshes, the number of iterations required by the linear solver to reach a converged solution is smaller. Furthermore, the results show that the strategy is applicable to coarse and fine meshes.

Table 3.2 collects all the data and the results. The number of iterations of the linear solver has been considered as a quality indicator of the analyses. As shown in the previous sections this value is related to the condition number of the iterative matrix.

Influence of the Time Step

The problem of Fig. 3.31 has been solved using different time steps: 0.0001, 0.0005, 0.001, 0.005, 0.01, 0.02 s. The mesh has a mean size of 0.15 m. The numerical results obtained with the reduced value of θ and by setting $\theta = 1$ are compared. Once again, the number of iterations of the linear solver is the parameter chosen to indicate the

Fig. 3.41 2D water sloshing. Number of iterations of the linear solver for different time step increments. Results for $\theta = 1$ and the optimum value of θ [33]

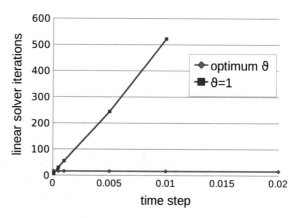

Table 3.3 2D water sloshing. Numerical values of the graph of Fig. 3.41 [33]

Δt (s)	Number of iterations	
	$\theta = 1$	Optimum θ
0.02	*Failed*	*16 ($\theta = 2.00 \cdot 10^{-5}$)*
0.01	*523*	*16 ($\theta = 8.01 \cdot 10^{-4}$)*
0.005	*244*	*16 ($\theta = 3.20 \cdot 10^{-3}$)*
0.001	*55*	*16 ($\theta = 8.01 \cdot 10^{-2}$)*
0.0005	*30*	*16 ($\theta = 3.89 \cdot 10^{-1}$)*
0.0001	*6*	*16 ($\theta = 8.01$)*

quality of the analyses: the smaller this value is, the better conditioned the linear system is.

The graph of Fig. 3.41 shows that the accuracy of the method does not depend on the time step increments, when the suitable reduced value for θ is used.

Table 3.3 summarizes the problem data and collects the numerical values of the graph of Fig. 3.41.

For each value of Δt, the number of iterations is 16 and the condition number does not change. Furthermore, using a reduced value of θ allows us to solve the problem for each time increment, while if θ is fixed to 1, the results are acceptable only until $\Delta t = 0.005$ s. For larger time steps the results for $\theta = 1$ are not accurate or the analyses do not even converge.

Mesh with a Refined Zone

As mentioned in the previous sections, the parameter θ has a square dependency on the mesh size (see Eq. (3.82)). For this reason, if the discretization is not uniform the global estimation of θ might not guarantee the well-conditioning of the linear system. In these cases, a local computation of θ is recommended. The sloshing problem is here solved again using the mesh shown in Fig. 3.42 and $\Delta t = 0.01$ s. The spatial discretization has two different mean sizes: in the center $h = 0.1$ m while $h = 0.4$ m elsewhere. The problem is solved both using $\theta = 1$ and by computing its optimum value globally and locally. The mean number of iterations required by the linear

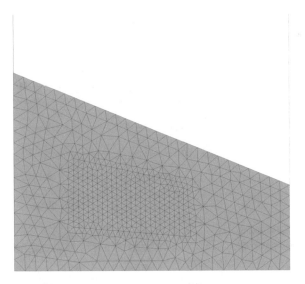

Fig. 3.42 2D water sloshing. Finite element mesh with a refined zone [33]

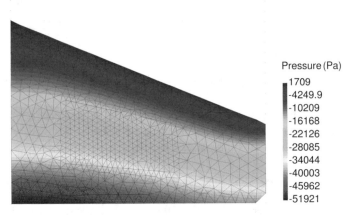

Pressure (Pa)
1709
-4249.9
-10209
-16168
-22126
-28085
-34044
-40003
-45962
-51921

Fig. 3.43 2D sloshing with a refined zone. Solution at time $t = 0.1$ s obtained with the local approach [33]

solver to converge for each of these three options is 319, 21 and 17 respectively. Hence, the local approach guarantees the best result for this type of non-uniform meshes.

In Fig. 3.43 the solution at time $t = 0.1$ s obtained with the local approach is illustrated.

Collapse of a Water Column on a Rigid Obstacle

In this section, the collapse of the water column induced by the instant removal of a vertical wall is studied.

Fig. 3.44 2D dam break. Initial geometry [33]

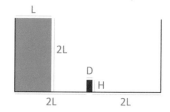

Table 3.4 2D dam break. Problem data

L	0.146 m
H	0.048 m
D	0.024 m
Viscosity	10^{-3} Pa · s
Density	10^3 kg/m^3
Bulk modulus	$2.15 \cdot 10^9$ Pa

The initial geometry of the problem is illustrated in Fig. 3.44. In Table 3.4 all the problem data are collected. As for the previous example, the problem is first solved with a very coarse mesh. Then a comparison with the solution obtained for $\theta = 1$ is given. After that, the same problem is solved varying the mean mesh size and for different time step increments. The results obtained with the optimum value of θ are compared to the experimental results presented in [57]. The objective is to show that the reduction of the bulk modulus in the iteration matrix does not affect the numerical solution of this class of impact problems which can be solved also with a larger time step.

The problem is first solved with a coarse discretization (mean element size $h = 0.0125$ m). The solutions obtained with the optimum value of θ and with $\theta = 1$ are compared in terms of the condition number of the iteration matrix and the number of iterations of the linear solver. For $\Delta t = 10^{-4}$ s, matrix K has condition numbers $C = 1028$ and $C = 60$, for $\theta = 1$ and $\theta = 0.0535$, respectively, which correspond to 1251 and 14 iterations of the linear solver, respectively. For the same discretization and a time increment of $\Delta t = 10^{-3}$ s, Eq. (3.81) yields $\theta = 0.000535$. The condition number of K and the iterations of the linear solver are the same as using a time step increment ten times smaller. Conversely, for $\theta = 1$, the condition number grows to $C = 102090$ and the iterative scheme does not converge.

Influence of the Mesh Size

The problem of Fig. 3.44 has been solved for $\Delta t = 10^{-4}$ s using unstructured meshes with the following mean element sizes: $h = 0.004, 0.005, 00075, 0.01, 0.0125$ m. The problem was solved both setting $\theta = 1$ and computing a priori its optimum value using Eq. (3.81). As for the previous section, the number of iterations of the linear solver has been considered as the quality indicator of the analysis. In Fig. 3.45 and Table 3.5 all the data and the results are collected.

Fig. 3.45 2D dam break. Number of iterations of the linear solver for different numbers of velocity degrees of freedom. Results for $\theta = 1$ and the optimum value of θ [33]

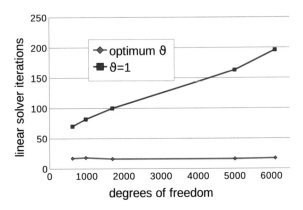

Table 3.5 2D dam break. Numerical values of the graph of Fig. 3.45 [33]

Average mesh size	Degrees of freedom (velocities)	Number of iterations	
		$\theta = 1$	Optimum θ
0.0125	618	70	17 ($\theta = 0.0536$)
0.01	978	82	18 ($\theta = 0.0345$)
0.0075	1694	100	16 ($\theta = 0.0205$)
0.005	5002	163	16 ($\theta = 0.00868$)
0.004	6106	196	17 ($\theta = 0.00559$)

Figure 3.45 and Table 3.5 confirm that the efficiency of the method is not affected by the mesh size. In other words, the well-conditioning of the iterative matrix is guaranteed for coarse and fine meshes.

Influence of the Time Step

The same problem was solved for different time steps: $\Delta t = 0.00005, 0.0001, 0.0005, 0.001, 0.0025$ s. The mesh used has a mean size of $h = 0.005$ m. Table 3.6 collects the problem data and the resulting number of iterations for all the analyses solved with $\theta = 1$ and with the optimum value of θ.

Figure 3.46 and Table 3.6 confirm the conclusions of the previous section.

The strategy is not affected by the time step and, contrary to the case for $\theta = 1$, it does not impose limitations to the range of time step increments for solving the problem. For example, for the present problem the time step increment is constrained by the geometry and the dynamics of the problem only. In other words, the maximum time step increment is the one that guarantees that the fluid particles do not cross through the boundaries. For the present problem and the chosen mesh, the maximum time step increment was $\Delta t = 2.5 \cdot 10^{-3}$ s.

Fig. 3.46 2D dam break. Number of iterations of the linear solver for different time steps. Results for $\theta = 1$ and the optimum value of θ [33]

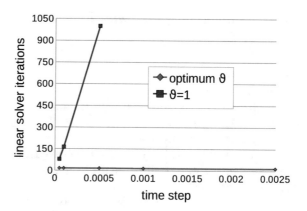

Table 3.6 2D dam break. Numerical values for the graph of Fig. 3.46 [33]

Δt (s)	Number of iterations	
	$\theta = 1$	Optimum θ
$2.5 \cdot 10^{-3}$	Failed	$15 \ (\theta = 1.39 \cdot 10^{-5})$
$1.0 \cdot 10^{-3}$	Failed	$15 \ (\theta = 8.68 \cdot 10^{-5})$
$5.0 \cdot 10^{-4}$	1000	$16 \ (\theta = 3.47 \cdot 10^{-4})$
$1.0 \cdot 10^{-4}$	163	$16 \ (\theta = 8.68 \cdot 10^{-3})$
$5.0 \cdot 10^{-5}$	79	$16 \ (\theta = 3.47 \cdot 10^{-2})$

All the examples presented in this section have shown the positive effect for the accuracy of the numerical scheme given by the use of the predicted value of κ_p in matrix K (Eq. (3.73)). In all the following problems presented in this work, this strategy has been used. In particular, the technique is still available in FSI problems for improving the conditioning of fluid counterpart of the global tangent matrix.

3.5 Validation Examples

This section is devoted to the validation of the unified stabilized formulation for fluids and solids at the incompressible limit. For each example several discretizations are analyzed in order to verify the convergence of the method. The formulation will be validated by comparing the numerical results to experimental tests and the numerical results of other formulations. In the first part quasi-incompressible Newtonian fluids are analyzed, then a problem involving a hypoelastic quasi-incompressible structure is studied.

3.5.1 Validation of the Unified Formulation for Newtonian Fluids

The purpose of this section is to validate the Unified formulation for the analysis of Newtonian free surface flows.

In all the examples presented in this section, in matrix K^κ (Eq. (3.77)) the pseudo bulk modulus κ_p is considered and its value has been computed according to the strategy presented in Sect. 3.4.3.2.

The effect of the boundary conditions is studied in detail. In particular, the strategy for modeling the slip conditions in the Lagrangian way described in Sect. 3.4.1.3 is tested and validated.

All the numerical examples have been solved for several meshes in order to verify the convergence of PFEM stabilized formulation.

Sloshing of a Viscous Fluid

In this section, the sloshing of a viscous fluid in a prismatic tank is analyzed. The initial configuration of the problem is illustrated in Fig. 3.47 and the problem data are given in Table 3.7.

Slip conditions have been imposed on the tank walls. The problem has been solved in 2D with different discretizations in order to verify the convergence of the numerical scheme. In particular, the following average mesh sizes of 3-noded triangles have been used: 0.012, 0.011, 0.01, 0.009, 0.008, 0.007, 0.006 and 0.005 m. In Fig. 3.48 the finest (mesh size $= 0.005$ m) and the coarsest (mesh size $= 0.012$ m) meshes are illustrated.

All the analyses have been run for $\Delta t = 0.001$ s and the total duration of the study is 10 s.

Fig. 3.47 Sloshing of a viscous fluid. Initial geometry of 2D simulation

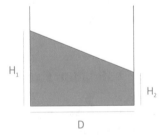

Table 3.7 Sloshing of a viscous fluid. Problem data		
D	0.5 m
H_1	0.35 m
H_2	0.15 m
Viscosity	5 Pa · s
Density	10^3 kg/m^3
Bulk modulus	$2.15 \cdot 10^9$ Pa

(a) average mesh size= 0.012m   **(b)** average mesh size= 0.005m

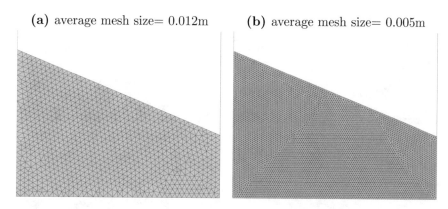

Fig. 3.48 Sloshing of a viscous fluid. Coarsest (1967 triangles) and finest (11440 triangles) meshes used for the analysis

In Fig. 3.49 some representative snapshots of the 2D simulation are given. Each snapshot refers to the maximum height reached by the fluid during its sloshing. The pressure contours have been plotted over the deformed configuration. The pictures in Fig. 3.49 assess the smoothness of the pressure field.

In Fig. 3.50 the time evolution of the fluid level height at the left side of the tank is plotted for the two finest meshes used. The graph shows that a converged solution has been reached because the two diagrams almost coincide.

The convergence analysis has been performed for the time evolution of the free surface position at the left side of the tank. For the convergence study, the solution obtained with the finest mesh (average size 0.005 m) has been considered as the reference solution. The error is computed as

$$
||err|| = \frac{\sqrt{\sum_{i=1}^{N} \left(h^i_{finest\ mesh} - h^i_{tested\ mesh} \right)^2}}{\sqrt{\sum_{i=1}^{N} \left(h^i_{finest\ mesh} \right)^2}}
\tag{3.85}
$$

where $h^i_{finest\ mesh}$ is the value computed for the finest mesh at the $i-th$ time step and N is the total number of time steps.

For a given mesh and for each time step the fluid height at the left side of the tank is compared to the one obtained with the finest mesh at the same time instant. The graph of Fig. 3.51 shows an overall quadratic slope, despite some oscillations in the curve. A reason for these oscillations is the local character of this convergence study. In order to reduce these drawbacks, a convergence study has been performed also for a global parameter, specifically the potential energy.

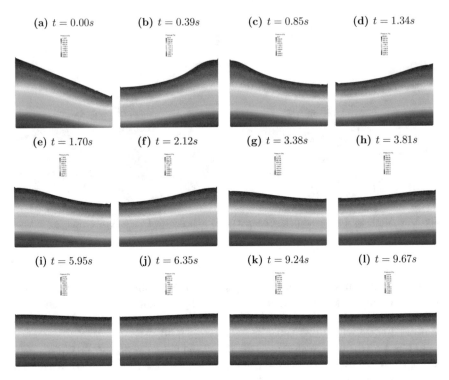

Fig. 3.49 Sloshing of a viscous fluid. Pressure contours over the deformed configuration at some time instants

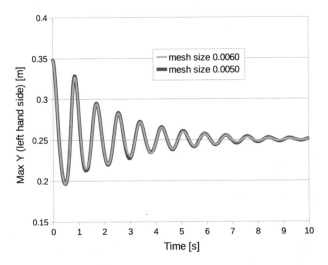

Fig. 3.50 Sloshing of a viscous fluid. Time evolution of the free surface position at the left side of the tank. Solutions obtained with the finest meshes: average mesh size 0.006 and 0.005 m

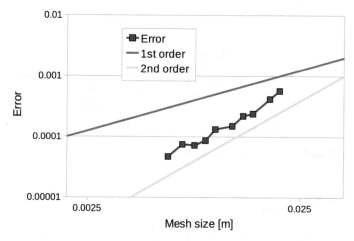

Fig. 3.51 Sloshing of a viscous fluid. Convergence analysis for the time evolution of the free surface level at the left side of the tank. Error computed with Eq. (3.85)

The potential energy has been computed as

$$E_{pot} = \sum_{e=1}^{n} m^e g h^e \qquad (3.86)$$

where m^e is the element mass and h^e is the mean value for the nodal heights.

The graph of Fig. 3.52 is the convergence curve for the potential energy error. Once again, the error is computed with respect to the results obtained with the finest mesh in the same form as in Eq. (3.85).

Comparing this graph to the curve of Fig. 3.51, the oscillations are reduced and the quadratic convergence for the used error measure is confirmed.

Collapse of a Water Column on a Rigid Horizontal Plane

In this section, the collapse of a water column on a rigid horizontal plane is studied. The initial configuration of the problem is illustrated in Fig. 3.53 and the problem data are given in Table 3.8.

The phenomena involved in this problem are the collapse of a free surface water column and the consequent spread of the water stream over a horizontal plate. The numerical solution is compared to the experimental results given in [58] where many experimental observations of the collapse of free liquid columns are collected. The characteristic variable chosen for the comparison is the residual height h of the water column (see Fig. 3.54).

Theproblem has been solved for both stick and slip conditions. The 2D simulation has been run for three different discretizations. In particular, the average mesh sizes 0.006, 0.0045 and 0.00075 m have been used. In Fig. 3.55 the finest (mesh size $= 0.00075$ m) and the coarsest (mesh size $= 0.006$ m) mesh are illustrated.

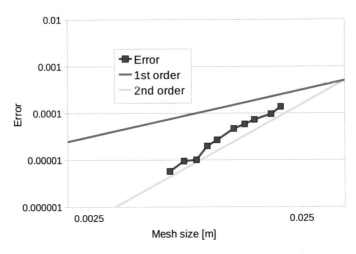

Fig. 3.52 Sloshing of a viscous fluid. Convergence graph for the potential energy. Error computed according to Eq. (3.85)

Fig. 3.53 Collapse of a water column. Initial geometry for the 2D simulation

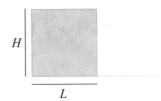

Table 3.8 Collapse of a water column. Problem data

H	0.05715 m
L	0.05715 m
Viscosity	10^{-3} Pa · s
Density	10^3 kg/m^3
Bulk modulus	$2.15 \cdot 10^9$ Pa

Fig. 3.54 Graphical definition of parameter h

In 3D the problem has been solved only for a mesh of 338839 4-noded tetrahedra with average size 0.002 m.

All the analyses have been run for $\Delta t = 0.00025$ s.

In Fig. 3.56 the velocity and the pressure fields obtained with the finest mesh and slip boundary conditions are plotted over the deformed configuration of the fluid.

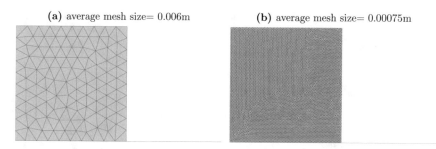

(a) average mesh size= 0.006m   **(b)** average mesh size= 0.00075m

Fig. 3.55 Collapse of a water column. Coarsest (258 triangles) and finest (13224 triangles) meshes used for the 2D analysis

The results for the 3D simulation with slip conditions are shown in Fig. 3.57.

In Fig. 3.58 the time evolution of the height at the left wall obtained with the finest mesh is compared to the values measured in the experimental tests [58].

The graph shows a very good agreement between the numerical and the experimental results.

For this problem, the solution obtained for slip and stick boundary conditions have been compared. In Fig. 3.59 the velocity contours obtained for both types of boundary conditions are plotted over the deformed configuration for some time instants. The mesh used is the coarsest one (average size 0.006 m).

In Fig. 3.60 the pressure field obtained for both boundary conditions is given.

The stick condition affects a layer that has a size of the same order of magnitude than the discretization. Hence, the coarser the mesh is, the bigger is the zone affected by the stick condition. When a coarse mesh is used, as in this case, imposing the stick condition penalizes excessively the motion and it can provoke the instability illustrated in the right column of Fig. 3.60. The stream obtained with the stick conditions is less uniform than the one obtained in the slip case and there are pressure concentrations. On the contrary, with a slip condition a fine solution is obtained also for the pressure field despite the coarseness of the mesh used in this example. This is because the motion of the wall particles helps in the remeshing step and in the global motion of the fluid stream.

In Fig. 3.61 the results for the velocity and the pressure fields obtained with the mesh with average size 0.00075 m and stick conditions are given. The pictures show that also for the stick case, a good solution for the pressure field can be obtained if the mesh is sufficiently fine.

The graph plotted in Fig. 3.62 shows that, also with the coarsest mesh tested in this example (average mesh size 0.006 m), the solution obtained with the slip condition is very close to the experimental results, while the numerical solution obtained for the same mesh but imposing stick conditions on the walls is quite far from the expected one.

However, as it is shown in the graphs of Fig. 3.63, the solution obtained with the stick condition tends to the slip solution with mesh refinement. In particular, using a discretization with an average mesh size of 0.00075 m the two solutions almost coincide.

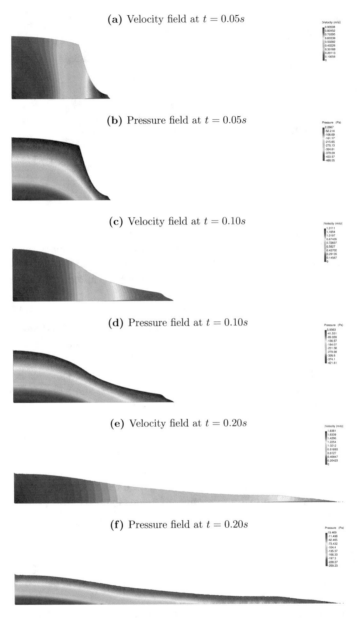

(a) Velocity field at $t = 0.05s$

(b) Pressure field at $t = 0.05s$

(c) Velocity field at $t = 0.10s$

(d) Pressure field at $t = 0.10s$

(e) Velocity field at $t = 0.20s$

(f) Pressure field at $t = 0.20s$

Fig. 3.56 Collapse of a water column. Pressure and velocity contours over the deformed configuration at three time instants

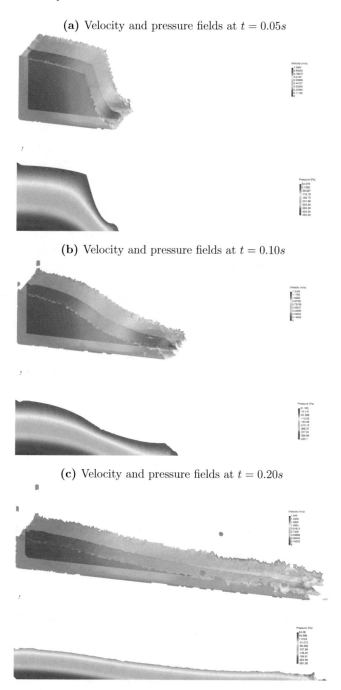

Fig. 3.57 Collapse of a water column. 3D simulation: velocity contours plotted over the 3D geometry and pressure contours drawn over the central section

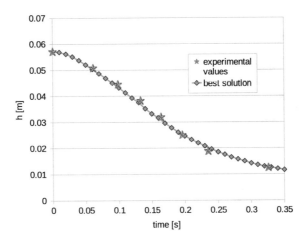

Fig. 3.58 Collapse of a water column: residual height h versus time. Comparison between the best numerical solution and the experimental results

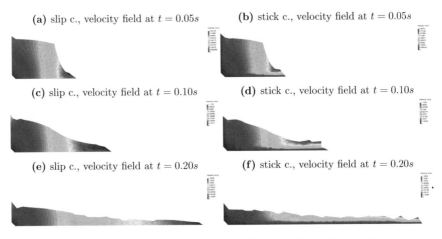

(a) slip c., velocity field at $t = 0.05s$ (b) stick c., velocity field at $t = 0.05s$

(c) slip c., velocity field at $t = 0.10s$ (d) stick c., velocity field at $t = 0.10s$

(e) slip c., velocity field at $t = 0.20s$ (f) stick c., velocity field at $t = 0.20s$

Fig. 3.59 Collapse of a water column. 2D results for slip and stick conditions and a coarse mesh (average size 0.006 m). Velocity contours over the deformed configuration at three time instants

Collapse of a Water Column Over a Rigid Step

The collapse of a water column over a rigid step is here studied in 2D and 3D. The initial configuration of the problem is illustrated in Fig. 3.44 and the problem data are given in Table 3.4.

The phenomena involved in this problem are the collapse of the water column, the impact against the rigid step, the subsequent creation of the wave, the impact against the vertical wall and the final mixing of the fluid. The numerical solution is compared to the experimental results given in [57], where many experimental observations of the collapse of free liquid columns are collected.

(a) slip c., pressure field at $t = 0.05s$ (b) stick c., pressure field at $t = 0.05s$

(c) slip c., pressure field at $t = 0.10s$ (d) stick c., pressure field at $t = 0.10s$

(e) slip c., pressure field at $t = 0.20s$ (f) stick c., pressure field at $t = 0.20s$

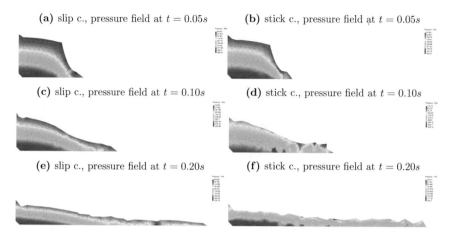

Fig. 3.60 Collapse of a water column. 2D results for slip and stick conditions and a coarse mesh (average size 0.006 m). Pressure contours over the deformed configuration at three time instants

(a) velocity field at $t = 0.05s$ (b) pressure field at $t = 0.05s$

(c) velocity field at $t = 0.10s$ (d) pressure field at $t = 0.10s$

(e) velocity field at $t = 0.20s$ (f) pressure field at $t = 0.20s$

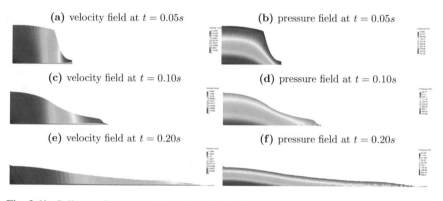

Fig. 3.61 Collapse of a water column. Results obtained with stick condition and a mesh with average size of 0.00075 m. Velocity and pressure contours over the deformed configuration at three time instants

The problem has been solved for both stick and slip conditions. A convergence study for different mesh sizes has been performed for the 2D simulation. In particular, for the convergence test the following average mesh sizes of 3-noded triangles have been used: 0.011, 0.0095, 0.009, 0.0085, 0.008, 0.0075, 0.007, 0.0065, 0.006, 0.0055, 0.005, 0.0045, 0.004, 0.003, 0.0025 and 0.002 m. In Fig. 3.64 the finest (mesh size $= 0.0002$ m) and the coarsest (mesh size $= 0.0011$ m) mesh are illustrated.

All the analyses have been run for $\Delta t = 0.0002$ s.

In Fig. 3.65 the numerical results obtained for the 2D simulation with the finest mesh are compared to the experimental observations [57] for the same instants. The velocity contours are plotted over the deformed configurations.

Fig. 3.62 Collapse of a
water column. Results for
slip and stick conditions for
the coarsest tested mesh
(average size 0.006 m). Time
evolution of the residual
height h of the water column

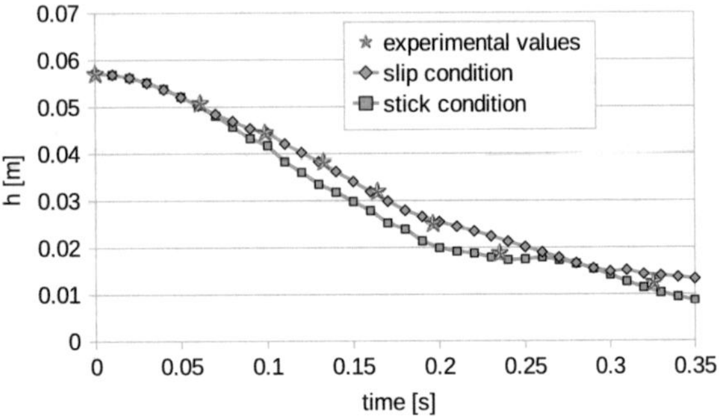

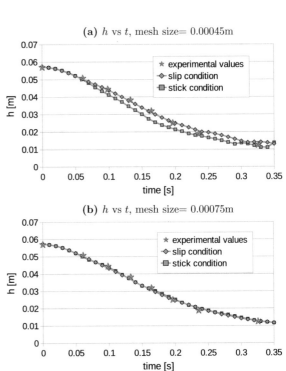

(a) h vs t, mesh size= 0.00045m

Fig. 3.63 Collapse of a
water column. Results for
slip and stick conditions for
different average mesh sizes.
Time evolution of the
residual height h of the water
column

(b) h vs t, mesh size= 0.00075m

The 3D problem has been solved considering a width of 0.146 m and an average
size of 0.008 m for the tetrahedra of the FEM mesh. This gives an initial mesh of
170711 4-noded tetrahedra. In Fig. 3.66 the 2D results (green) are superposed to the
cut performed at the center of the 3D domain (grey) for the same time instants. An
average size of 0.008 m has been used for the edges of both the triangles and the
tetrahedra for the 2D and 3D discretizations.

The Fig. 3.67 refers to the slip problem only and it shows that the solutions obtained
with the three finest meshes are very similar. This is a further evidence of the convergence of the PFEM strategy here presented.

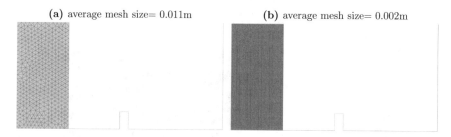

Fig. 3.64 Collapse of a water column over a rigid step. Coarsest (1014 triangles) and finest (24668 triangles) meshes used for the 2D analysis

In Fig. 3.68 the solution obtained at $t = 0.3$ s with a stick condition and the finest mesh is given. The differences between the streams are very small.

As already mentioned, the convergence analysis has been performed for both cases of slip and stick conditions imposed on the walls. Specifically, the study of convergence has been performed first for the maximum velocity and then the kinetic energy of the whole domain. Both values have been computed at $t = 0.1$ s, so just before the impact with the rigid step.

In the graph of Fig. 3.69 the Y-axis is the maximum velocity obtained for the stick and slip conditions and the X-axis refers to the degrees of freedom (dof) for the velocity. The solutions obtained with the stick condition suffer from oscillations for the coarsest meshes, but for finer meshes tend to those of the slip condition.

A convergence study for the kinetic energy of the whole domain at $t = 0.1$ s has been performed. The kinetic energy has been computed as

$$E_{kin} = \sum_{e=1}^{n} 0.5 m^e (v^e)^2 \tag{3.87}$$

where m^e is the element mass and v^e is the mean value for the nodal velocities of the element.

The graph of Fig. 3.70 is the evolution of the kinetic energy for the first 0.1 s of analysis, for both cases of slip and stick conditions and for three different meshes: the coarsest one (mesh size $= 0.011$ m), the finest one (mesh size $= 0.002$ m) and an intermediate discretization (mesh size $= 0.006$ m). The curves for the slip case are almost superposed, contrary to the ones for the stick problem. However, for the finest mesh, the curves for the slip and stick conditions are very close.

In the graph of Fig. 3.71 the Y-axis refers to the kinetic energy at $t = 0.1$ s for the stick and slip conditions and the X-axis refers to the degrees of freedom for the velocity. With slip conditions, the kinetic energy has almost the same value for all the discretizations (the difference between the values given by the coarsest and the finest discretization is lower than 0.5 %). However for stick conditions the kinetic

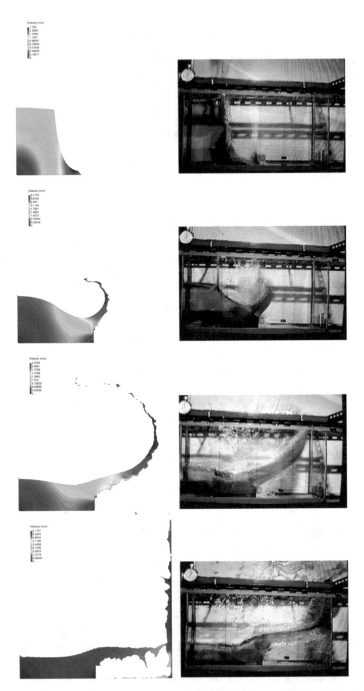

Fig. 3.65 Collapse of a water column over a rigid step. Velocity contours over the deformed configuration at time: $t = 0.1, 0.2, t = 0.3$ and 0.4 s. Results obtained with the finest mesh and slip conditions on the walls

Fig. 3.66 Collapse of a water column over a rigid step. Superposition of the results obtained with an average mesh size of 0.008 m for the 2D and 3D simulations (*green* and *grey* colours, respectively) and with slip conditions on the walls

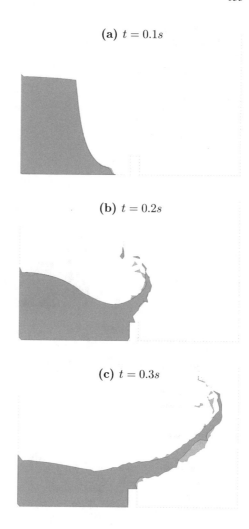

(a) $t = 0.1s$

(b) $t = 0.2s$

(c) $t = 0.3s$

energy grows with the refinement of the mesh and it seems to converge to the value of the slip case.

The convergence curve for the kinetic energy evolution on time for the first 0.1 s of analysis is given in Fig. 3.72. For both slip and stick cases, the error has been computed using to Eq. (3.85). The slope is almost 2 for both curves.

In Fig. 3.73 the time evolution of the pressure until $t = 0.1$ s is studied. The graph on the left refers to a coarse mesh and the one in the right to the finest mesh. The plots show that the boundary conditions affect the pressure field. In particular the stick condition induces oscillations in the pressure solution which amplitude reduces by refining the mesh.

This analysis has shown that for coarse meshes there are significant differences between the solution given by imposing stick or slip conditions. However, for fine meshes the problems solved with either slip or stick conditions tend to the same

Fig. 3.67 Collapse of a
water column over a rigid
step. Superposition of the 2D
results obtained at $t = 0.3$ s
with the meshes with
average size 0.002 m
(*orange*), 0.0025 m (*pink*)
and 0.003 m (*green*) (slip
condition on the walls)

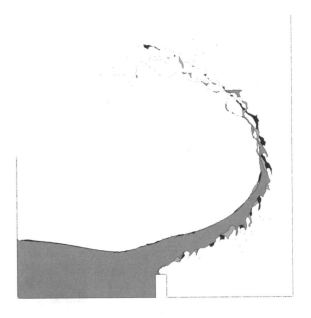

Fig. 3.68 Collapse of a
water column over a rigid
step. Superposition of the 2D
results obtained at $t = 0.3$ s
with the mesh with average
size 0.002 m for the slip
(*orange colour*) and the stick
(*pink*) conditions

solution. In fact, for coarse meshes, the stick conditions affect highly the motion
of the fluid and a refinement is required in order to obtain a solution close to the
expected one. Instead, with slip conditions on the walls it is possible to obtain fine
results, for both the velocity and pressure fields, also for coarse meshes.

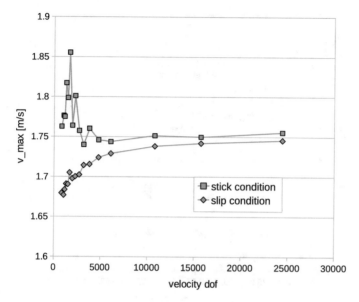

Fig. 3.69 Collapse of a water column over a rigid step. Maximum velocity at $t = 0.1$ s for slip and stick conditions (2D simulation)

3.5.2 Validation of the Unified Formulation for Quasi-incompressible Hypoelastic Solids

The Unified formulation is here validated for quasi-incompressible solids by solving a benchmark problem for non-linear solid mechanics, the Cook's membrane. In Sect. 2.3.4 the same structure was analyzed considering a compressible material. In this section, the nearly-incompressible problem is analyzed. In particular, first the problem is solved for $\nu = 0.4999$ with all the solid elements derived in this thesis, the V, the VP and the VPS elements comparing the numerical results for the displacements, the pressure and the stresses. Then the maximum displacement given by the three elements solving the same problem for the range of Poisson ratio from 0.3 to 0.499999 is compared.

Nearly Incompressible Cook's Membrane

In Fig. 3.74a the initial geometry and the problem data are given.

The problem has been solved in 2D for different discretizations in order to verify the convergence of the scheme. In this case, unstructured meshes have been used. In Fig. 3.74b the coarsest finite element discretization (average size 5) used in this example is depicted. The finest mesh tested in this problem has mean size 0.25 that gives a mesh of 52372 triangles.

As for the previous case, the problem is studied in statics and the self-weight of the membrane has not been taken into account. The reference solution for this problem is

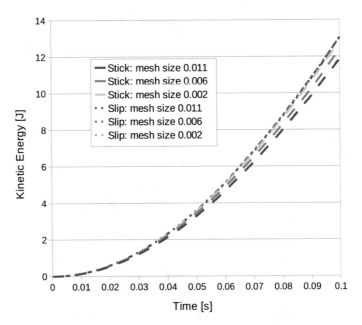

Fig. 3.70 Collapse of a water column over a rigid step. Time evolution of the kinetic energy for slip and stick conditions and for three different FEM discretizations (2D simulation)

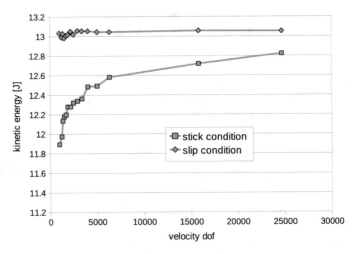

Fig. 3.71 Collapse of a water column over a rigid step. Kinetic energy at $t = 0.1$ s for slip and stick conditions

taken from [59], where the problem was solved with an Incompatible Bubble element. In this publication the tip vertical displacement obtained U_Y^{max} by other formulations is also given. Specifically the results are provided for a FEM formulation with linear displacements and constant pressure and the Enhanced Assumed Strain formulation

Fig. 3.72 Collapse of a
water column over a rigid
step. Convergence analysis
for the kinetic energy
evolution on time for the first
0.1 s of analysis for slip and
stick conditions (2D case)

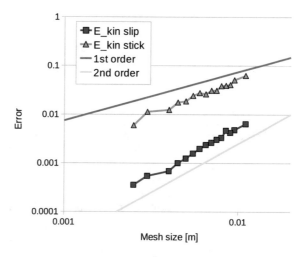

Fig. 3.73 Collapse of a
water column over a rigid
step. Time evolution of the
pressure for a mesh with
average size 0.006 m (*top*)
and 0.002 m (*bottom*).
Solution for stick and slip
conditions (2D simulation)

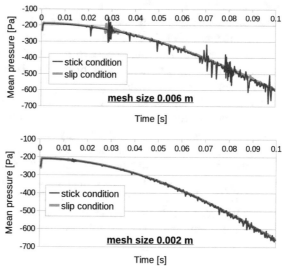

[60, 61]. For all these formulations and the finest mesh tested in [59], this value is around $U_Y^{max} = 7.71$.

The Poisson ratio 0.4999 represents a material that is almost incompressible. It is well known that values of the Poisson ratio close to 0.5 generate numerical problems to non-stabilized FEM schemes, such as the locking of the solution. Apart from this, the proximity to the incompressible limit also produces ill-conditioned matrices and deteriorates the convergence of the solution. In order to overcome these drawbacks a stabilized formulation is required. In this work, the stabilization of the quasi incompressible mixed formulation for the solid element is achieved via the FIC strategy presented in Sect. 3.1 using the VPS-element. The quasi-incompressible

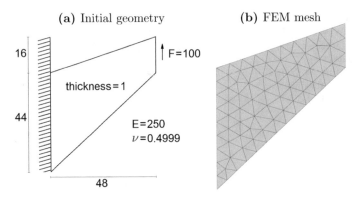

Fig. 3.74 Nearly incompressible Cook's membrane. Initial geometry, material data and FEM mesh (127 elements)

membrane is here also solved using the non-stabilized V and VP elements derived in Chap. 2. Both these elements, although in different ways, suffer from instability near the incompressible limit of the material. Despite that, this example is presented with the purpose of showing how each formulation is affected by the mentioned drawbacks and to underline the superiority of a mixed formulation for dealing with nearly incompressible materials.

In Fig. 3.75 the top corner vertical displacements obtained with the V, the VP and the VPS elements for different FEM meshes are plotted.

In Table 3.9 the numerical values are given.

For the V, VP and VPS the percentage errors for the tip displacement with respect to the reference solution are 5.68, 0.707 and 0.791 %, respectively.

For the analysis of stress results, the case of the mesh with average size 1.5 is studied. In Fig. 3.76 the X-component of the Cauchy stress tensor obtained with V, VP, and VPS elements is plotted over the deformed configurations. From the pictures it is clear that both the V and VP solutions deteriorate for values of ν close to the

Fig. 3.75 Nearly incompressible Cook's membrane. Top corner vertical displacement for the V, VP and the VPS elements

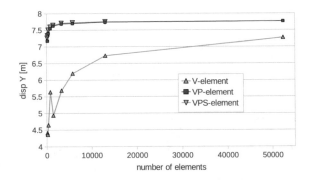

Table 3.9 Nearly incompressible Cook's membrane. Top corner vertical displacement for different formulations and discretizations [62]

Mesh size	Number of elements	V-element	VP-element	VPS-element
		U_y	U_y	U_y
5	127	4.411	7.031	7.268
4	194	4.365	7.178	7.338
3	361	4.648	7.401	7.508
2	802	5.643	7.551	7.603
1.5	1441	4.937	7.632	6.655
1	3288	5.690	7.695	7.714
0.75	5806	6.199	7.707	7.729
0.5	13015	6.731	7.745	7.755
0.25	52372	7.272	7.765	7.771

(a) V-element   (b) VP-element   (c) VPS-element

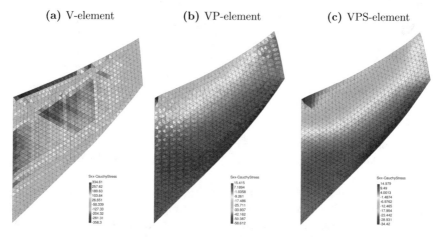

Fig. 3.76 Nearly incompressible Cook's membrane. Results of the XX component of the Cauchy stress tensor for V, VP and VPS elements [62]

incompressible limit of the material, while the stress field given by the VPS-element is good and smooth.

In Fig. 3.77 the pressure solution obtained with the VP-element and the VPS-element are given.

The non-stabilized VP-element yields a pressure field that is completely untrustworthy exhibiting the classical checkerboard modes. Instead, the solution of the VPS element is smooth and accurate.

From the kinematic point of view, the displacements obtained by the mixed formulation (VP and VPS elements) are close to the expected solution, while the solution given by the V-element is totally locked. For the mesh displayed in Figs. 3.76 and 3.77

(a) VP-element   **(b)** VPS-element

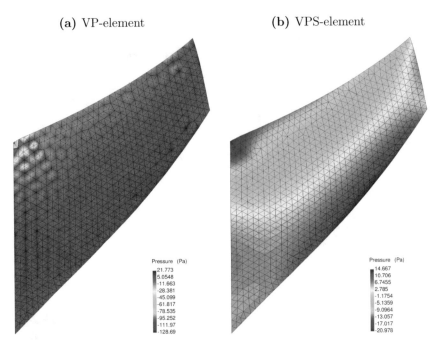

Fig. 3.77 Nearly incompressible Cook's membrane. Pressure solution for the VP and VPS elements [62]

(average mesh size 1.5) the errors for the top corner displacement with respect the reference solution are 36.5, 1.9 and 1.61 % for the V, the VP and the VPS elements, respectively.

The same problem is solved for different Poisson ratios, from 0.3 to 0.499999, using the same FEM discretization (average mesh size equal to 1). For the stabilized formulation the Poisson ratio that appears in the tangent matrix of the linear momentum equations has been limited to 0.4999. Instead, in the equation of the pressure the actual value of the Poisson ratio is used. This technique is similar to the one used for the stabilization of the fluid in Sect. 3.4.3 where the bulk modulus is reduced only in the tangent matrix of the linear momentum equations but not in the continuity equation. The graphs of Fig. 3.78 are the plots of the top corner vertical displacement obtained for the different Poisson ratios using the V, the VP and the VPS elements (the graph at the right is a zoom of the last part of the graph at the left).

In Table 3.10 the numerical values are given.

The graphs show that for Poisson ratios bigger than 0.45, the velocity formulation exhibits locking. Instead, the mixed formulations despite the increasing of ill-conditioning of the linear system, yields a good solution until a Poisson ratio of 0.49999. Beyond this value, the non-linear solver does not even converge for the non-stabilized VP-element, while the solution given by the VPS-element is still a good one.

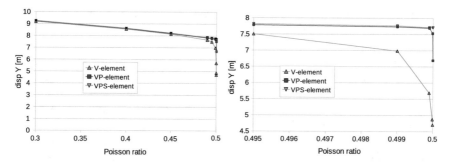

Fig. 3.78 Nearly incompressible Cook's membrane. Maximum vertical displacement for different Poisson ratios. Results for V, VP and VPS elements. The graph at the *right* is a zoom of the last part of the graph at the *left*

Table 3.10 Nearly incompressible Cook's membrane. Numerical values for the maximum vertical displacement for different Poisson ratios obtained with V, VP and VPS elements

Poisson ratio	V-element	VP-element	VPS-element
0.3	9.176	9.212	9.245
0.4	8.557	8.601	8.628
0.45	8.143	8.203	8.230
0.49	7.658	7.833	7.866
0.495	7.503	7.781	7.817
0.499	6.988	7.735	7.766
0.4999	5.690	7.695	7.714
0.49999	4.872	7.532	7.714
0.499999	4.706	6.696	7.713

3.6 Summary and Conclusions

In this chapter the Unified Stabilized formulation for quasi-incompressible materials has been derived. This numerical procedure is based on the mixed Velocity–Pressure formulation derived in Chap. 2 for compressible materials.

In order to deal with material incompressibility using a linear interpolation for both the pressure and the velocity fields, the scheme required to be stabilized. The necessary stabilization has been introduced using an enhanced version of the Finite Calculus (FIC) method. In Sect. 3.1 the complete derivation of the stabilized form of the mass balance equation has been presented. The FIC stabilization has been derived for quasi-incompressible fluids and has been extended also to hypoelastic quasi-incompressible solids. The solid finite element based on the mixed Velocity–Pressure stabilized formulation has been called VPS element.

The solution schemes for solving quasi-incompressible Newtonian fluids and hypoelastic solids with the stabilized mixed Velocity–Pressure formulation have been given and explained in detail.

Section 3.4 has been fully devoted to the analysis of free surface fluids with the Unified stabilized formulation. First the Particle Finite Element Method (PFEM) has been explained and its advantages and disadvantages have been highlighted. A simple technique for modeling the slip conditions moving the wall particles have been also given. Then mass preservation properties of the PFEM-FIC stabilized formulation have been tested with several numerical examples. It has been shown that the method yields excellent results for a variety of 2D and 3D free surface flow problems involving surface waves, water splashing, violent impact of flows with containment walls and mixing of fluids. Next the conditioning of the scheme has been studied and the numerical inconveniences associated to the high value of the bulk modulus have been highlighted. An efficient and easy to implement technique for improving the conditioning and the global convergence of the problem using a scaled value for the pseudo bulk modulus has been given. The strategy has been validated for two benchmark problems for free surface flows.

In the last section of this chapter several numerical examples have been presented for validating the unified stabilized formulation for solving fluids and solids close to the incompressible limit. The proposed method has been validated versus both experimental tests and numerical results from other formulations and it has been shown that the method is convergent to the expected solution. Particular attention has been given to the analysis of the boundary conditions. In particular, it has been shown that for inviscid fluids and coarse meshes it is preferable to use slip conditions in order to avoid the pressure concentrations induced by the stick conditions.

References

1. F. Brezzi. On the existence, uniqueness and approximation of saddle-point problems arising from lagrange multipliers. *Revue française d'automatique, informatique, recherche opéra-tionnelle. Série rouge. Analyse numérique*, 8(R-2):129–151, 1974.
2. F. Felippa and E. Oñate. Nodally exact ritz discretizations of 1d diffusion-absorption and helmholtz equations by variational fic and modified equation methods. *Computational Mechanics*, 39:91–111, 2007.
3. E. Oñate. Derivation of stabilized equations for advective-diffusive transport and fluid flow problems. *Computer methods in applied mechanics and engineering*, 151:233–267, 1998.
4. E. Oñate, J. Rojek, R.L. Taylor, and O.C. Zienkiewicz. Finite calculus formulation for incompressble solids using linear triangles and tetrahedra. *International Journal For Numerical Methods In Engineering*, 59:1473–1500, 2004.
5. E. Oñate, A. Valls, and J. García. Fic/fem formulation with matrix stabilizing terms for incompressible flows at low and high reynold's numbers. *Computational mechanics*, 38 (4–5):440–455, 2006.
6. E. Oñate, A. Valls, and J.García. Computation of turbulent flows using a finite calculus-finite element formulation. *International Journal of Numerical Methods in Engineering*, 54:609–637, 2007.

7. E. Oñate, P. Nadukandi, S.R. Idelsohn, J. García, and C. Felippa. A family of residual-based stabilized finite element methods for stokes flows. *International Journal for Numerical Methods in Fluids*, 65 (1–3):106–134, 2011.

8. E. Oñate, Idelsohn SR, and Felippa C. Consistent pressure laplacian stabilization for incompressible continua via higher-order finite calculus. *International Journal of Numerical Methods in Engineering*, 87 (1–5):171–195, 2011.

9. E. Oñate, A. Franci, and J.M. Carbonell. Lagrangian formulation for finite element analysis of quasi-incompressible fluids with reduced mass losses. *International Journal for Numerical Methods in Fluids*, 74 (10):699–731, 2014.

10. E. Oñate, S.R. Idelsohn, F. Del Pin, and R. Aubry. The particle finite element method. an overview. *International Journal for Computational Methods*, 1:267–307, 2004.

11. H. Edelsbrunner and E.P. Mucke. Three dimensional alpha shapes. *ACM Trans Graphics*, 13:43–72, 1999.

12. T. Belytschko, W.K. Liu, B. Moran, and K.I. Elkhodadry. *Nonlinear Finite Elements For Continua And Structures. Second Edition*. John Wiley & Sons, New York, 2014.

13. J. Donea and A. Huerta. *Finite Element Methods for Flow Problems*. Wiley, 2003.

14. O.C. Zienkiewicz and R.L. Taylor. *The Finite Element Method for Solid and Structural Mechanics, Volume 2 (6th Ed.)*. Elsiever Butterworth-Heinemann, Oxford, 2005.

15. O.C. Zienkiewicz, R.L. Taylor, and P. Nithiarasu. *The Finite Element Method for Fluid Dynamics, Volume 3 (6th Ed.)*. Elsiever, Oxford, 2005.

16. S.R. Idelsohn and E. Oñate. The challenge of mass conservation in the solution of free surface flows with the fractional step method. problems and solutions. *Communications in Numerical Methods in Engineering*, 26 (10):1313–1330, 2008.

17. S.R. Idelsohn, M.Mier-Torrecilla, and E. Oñate. Multi-fluid flows with the particle finite element method. *Computer methods in applied mechanics and engineering*, 198:2750–2767, 2009.

18. A. Limache, S.R. Idelsohn, R. Rossi, and E. Oñate. The violation of objectivity in laplace formulation of the navier-stokes equations. *International Journal for Numerical Methods in Fluids*, 54:639–664, 2007.

19. E. Turkel. Preconditioned methods for solving the incompressible and low speed compressible equations. *Journal of Computational Physics*, 72:277–298, 1987.

20. A.J. Chorin. A numerical method for solving incompressible viscous flow problems. *Journal of Computational Physics*, 135:118–125, 1997.

21. S.R. Idelsohn, J. Marti, A. Limache, and E. Oñate. Unified lagrangian formulation for elastic solids and incompressible fluids: Applications to fluid-structure interaction problems via the pfem. *Computer Methods In Applied Mechanics And Engineering*, 197:1762–1776, 2008.

22. E. Oñate and J. García. A finite element method for fluid-structure interaction with surface waves using a finite calculus formulation. *Computer methods in applied mechanics and engineering*, 191:635–660, 2001.

23. E. Oñate. Possibilities of finite calculus in computational mechanics. *International Journal of Numerical Methods in Engineering*, 60 (1):255–281, 2004.

24. E. Oñate. A stabilized finite element method for incompressible viscous flows using a finite increment calculus formulation. *Computer Methods in Applied Mechanics and Engineering*, 190 (20–21):355–370, 2000.

25. E. Oñate, M.A. Celigueta, S.R. Idelsohn, F. Salazar, and B. Suarez. Possibilities of the particle finite element method for fluid–soil–structure interaction problems. *Computation mechanics*, 48:307–318, 2011.

26. E. Oñate, A. Franci, and J.M. Carbonell. A particle finite element method for analysis of industrial forming processes. *Computational Mechanics*, 54:85–107, 2014.

27. E. Oñate, M.A. Celigueta, and S.R. Idelsohn. Modeling bed erosion in free surface flows by th particle finite element method. *Acta Geotechnia*, 1 (4):237–252, 2006.

28. E. Oñate and J.M. Carbonell. Updated lagrangian finite element formulation forquasi and fully incompressible fuids. *Computational Mechanics*, 54 (6), 2014.

29. E. Oñate, A. Franci, and J.M. Carbonell. A particle finite element method (pfem)for coupled thermal analysis of quasi and fully incompressible flows and fluid structure interaction

problems. *Numerical Simulations of Coupled Problems in Engineering. S.R. Idelsohn (Ed.)*, 33:129–156, 2014.

30. E. Oñate, J. García, S.R. Idelsohn, and F. Del Pin. Fic formulations for finite element analysis of incompressible flows. eulerian, ale and lagrangian approaches. *Computer methods in applied mechanics and engineering*, 195 (23–24):3001–3037, 2006.

31. P. Ryzhakov, E. Oñate, and S.R. Idelsohn. Improving mass conservation in simulation of incompressible flows. *International Journal of Numerical Methods in Engineering*, 90:1435–1451, 2012.

32. E. Oñate, P. Nadukandi, and S. Idelsohn. P1/p0+ elements for incompressible flows with discontinuous material properties. *Computer Methods in Applied Mechanics and Engineering*, 271:185–209, 2014.

33. A. Franci, E. Oñate, and J.M. Carbonell. On the effect of the bulk tangent matrix in partitioned solution schemes for nearly incompressible fluids. *International Journal for Numerical Methods in Engineering*, 102:257–277, 2015.

34. PFEM in CIMNE website. www.cimne.com/pfem.

35. S.R. Idelsohn, E. Oñate, and F. Del Pin. The particle finite element method: a powerful tool to solve incompressible flows with free-surfaces and breaking waves. *International Journal for Numerical Methods in Engineering*, 61:964–989, 2004.

36. A. Larese, R. Rossi, E. Oñate, and S.R. Idelsohn. Validation of the particle finite element method (pfem) for simulation of free surface flows. *International Journal for Computer-Aided Engineering and Software*, 25:385–425, 2008.

37. S. Idelsohn, N. Calvo, and E. Oñate. Polyhedrization of an arbitrary point set. *Computer Methods in Applied Mechanics and Engineering*, 92 (22–24):2649–2668, 2003.

38. A. Saalfeld. In *Delaunay edge refinements*, pages 33–36, Burnaby, 1991.

39. H. Edelsbrunner and T.S. Tan. An upper bound for conforming delaunay triangulations. *Discrete and Computational Geometry*, 10 (2):197–213, 1993.

40. D. Cohen-Steiner, E. Colin de Verdiere, and M. Yvinec. Conforming delaunay triangulations in 3d. *Special issue on the 18th annual symposium on computational Geometry*, 28 (2–3):217–233, 2004.

41. X. Zhang, K. Krabbenhoft, D.M. Pedroso, A.V. Lyamin, D. Sheng, M. Vicente da Silva, and D. Wang. Particle finite element analysis of large deformation and granular flow problems. *Computer and Geotechnics*, 54:133–142, 2013.

42. E. Oñate, S.R. Idelsohn, M.A. Celigueta, and R. Rossi. Advances in the particle finite element method for the analysis of fluid–multibody interaction and bed erosion in free surface flows. *Computer methods in applied mechanics and engineering*, 197 (19–20):1777–1800, 2008.

43. D. Mavriplis. Advancing front delaunay triangulation algorithm designed for robustness. *Journal of Computational Physics*, 117:90–101, 1995.

44. J.M. Carbonell, E. Oñate, and B. Suarez. Modeling of ground excavation with the particle finite-element method. *Journal of Engineering Mechanics*, 136:455–463, 2010.

45. J.M. Carbonell, E. Oñate, and B. Suarez. Modelling of tunnelling processes and cutting tool wear with the particle finite element method (pfem). *Computational Mechanics*, 52 (3):607–629, 2013.

46. E. Oñate, R. Rossi, S.R. Idelsohn, and K. Butler. Melting and spread of polymers in fire with the particle finite element method. *International Journal of Numerical Methods in Engineering*, 81 (8):1046–1072, 2010.

47. X. Oliver, J.C. Cante, R. Weyler, C. González, and J. Hernández. *Particle finite element methods in solid mechanics problems*. In: Oñate E, Owen R (Eds) Computational Plasticity. Springer, Berlin, 2007.

48. J.M. Carbonell. *Doctoral thesis: Modeling of Ground Excavation with the Particle Finite Element Method*. 2009.

49. M. Cremonesi and U. Perego. Numerical simulation of landslide-reservoir interaction using a pfem approach. In *Proceedings of Particle-Based Methods III. Fundamentals and Applications*, pages 408–417, Stuttgardt, 2013.

50. M. Cremonesi, F. Ferri, and U. Perego. A lagrangian finite element approach for the numerical simulation of landslide runouts. In *XX Italian Conference on Computational Mechanics, VII Italian Meeting of AIMETA Material Group, Ed. by Sacco and S. Marfia*, pages 1–2, Cassino, 2014.

51. E. Oñate, J. Rojek, S.R. Idelsohn, F. Del Pin, and R. Aubry. Advances in syabilized finite element and particle methods for bulk forming processes. *Computer methods in applied mechanics and engineering*, 195:6750–6777, 2006.

52. E. Oñate, J. Marti, R. Rossi, and S.R. Idelsohn. Analysis of the melting, burning and flame spread of polymers with the particle finite element method. *Computer Assisted Methods in Engineering and Science*, 20:165–184, 2013.

53. P. Ryzhakov, J. Cotela, R. Rossi, and E. Oñate. A two-step monolithic method for the efficient simulation of incompressible flows. *International Journal for Numerical Methods in Fluids*, 74 (12):919–934, 2014.

54. P. Ryzhakov, R. Rossi, S.R. Idelsohn, and E. Oñate. A monolithic lagrangian approach for fluid-structure interaction problems. *Computational Mechanics*, 46:883–899, 2010.

55. K.J. Bathe. *Finite Element Procedures*. Prentice-Hall, New Jersey, 1996.

56. B. Hubner, E. Walhorn, and D. Dinkler. A monolithic approach to fluid-structure interaction using space-time finite elements. *Computer Methods in Applied Mechanics and Engineering*, 193:2087–2104, 2004.

57. D.M. Greaves. Simulation of viscous water column collapse using adapting hierarchical grids. *International Journal of Numerical Methods in Engineering*, 50:693–711, 2006.

58. J.C. Martin and W.J. Moyce. Part iv. an experimental study of the collapse of liquid columns on a rigid horizontal plane. *Philosophical Transactions of the Royal Society of London. Series A, Mathematical and Physical Sciences*, 244 (882):312–324, 1952.

59. I. Romero and M. Bishoff. Incompatible bubbles: A non-conforming finite element formulation for linear elasticity. *Computer Methods In Applied Mechanics And Engineering*, 196:1662–1672, 2006.

60. J.C. Simo and M.S. Rifai. A class of mixed assumed strain methods and the method of incompatible modes. *International Journal for Numerical Methods in Engineering*, 29(8):1595–1638, 1990.

61. M. Bishoff and I. Romero. A generalization of the method of incompatible modes. *International Journal for Numerical Methods in Engineering*, 69 (9):1851–1868, 2007.

62. A. Franci, E. Oñate, and J. M. Carbonell. Velocity-based formulations for standard and quasi-incompressible hypoelastic-plastic solids. *International Journal for Numerical Methods in Engineering*, doi:10.1002/nme.5205, 2016.

Chapter 4
Unified Formulation for FSI Problems

4.1 Introduction

In the previous chapters the Unified formulation was used for solving the mechanics of fluids and solids separately. In this chapter it will be shown that the method is also suitable for solving fluid–structure interaction (FSI) problems. The algorithm for coupling the three solid formulations (the V, the VP and the VPS elements) with the mixed Velocity–Pressure FIC-stabilized formulation for Newtonian fluids is described. It will be shown that the proposed method gives the possibility to select the formulation for the solid depending on the problem to solve. The numerical results will be validated by comparing the solution of benchmark FSI problems with the results of the literature.

In this work, the FSI problems are solved in a monolithic way. This means that fluids and solids are solved within the same linear system and the iterations between the fluid and the solid solver are not required, as for staggered schemes. The Unified formulation uses the same framework (Lagrangian description), the same spatial discretization and the same temporal integration schemes (linear shape functions and implicit integration of time), the same unknown variables (velocities and pressures), and the same solution method (two-step partitioned scheme) for both fluid and solid mechanics solution. In conclusion, solids and fluids just represent different regions of the same global domain and they differ on the characteristic material parameters only. All this simplifies the implementation of the FSI solver. In fact, it is not required any variable transformation and the implementation effort for coupling the mechanics of fluids and solids is reduced to a proper assembly of the global linear system.

Concerning the mesh, there must be a correspondence at the interface between the solid and the fluid nodes, in the spirit of conforming mesh methods. It will be shown that the PFEM guarantees the conformity.

In conclusion, for explaining the extension of the Unified formulation to FSI problems only the assembly of the global linear system and the way to detect the fluid–solid interface have to be described. This will be done in the next section. Then

© Springer International Publishing AG 2017

A. Franci, *Unified Lagrangian Formulation for Fluid and Solid Mechanics,*
Fluid-Structure Interaction and Coupled Thermal Problems Using the PFEM,
Springer Theses, DOI 10.1007/978-3-319-45662-1_4

the solution schemes for solving FSI problems coupling the FIC-stabilized mixed VP formulation for quasi-incompressible Newtonian fluids with the V, the VP and the VPS elements for hypoelastic solids are given. Finally, several numerical examples of benchmark FSI problems are solved and all the schemes are validated and compared.

4.2   FSI Algorithm

The assembly of the global linear system is performed by making a loop over all the nodes of the mesh. Each node provides the contributions of the elements that share the node, and each element is computed according to the specific constitutive law and solution scheme. So, when an interface node is analyzed, it is necessary to sum the contributions of both materials in the global linear system (see Fig. 4.1 for a graphic representation of the global assembly).

Those nodes that belong to a single domain assemble to the global matrix with the elemental contributions of one material only.

Concerning the degrees of freedom, each node of the mesh is characterized by a single set of kinematic variables. So they move according to a unique velocity. This means that the degrees of freedom for the solid and fluid velocities coincide also at the interface nodes. On the other hand, in order to guarantee the correct boundary conditions for the stresses, each interface node has a degree of freedom for the pressure of the fluid and, for the VP and the VPS elements, for the pressure of the solid. This means that, for each interface node the fluid elements assemble only the contributions for the degree of freedom of the fluid pressure, while the solid elements do that for the solid pressure. This requires solving twice the continuity equation: once for the fluid domain and once for the VP-element or the VPS-element for the solid domain.

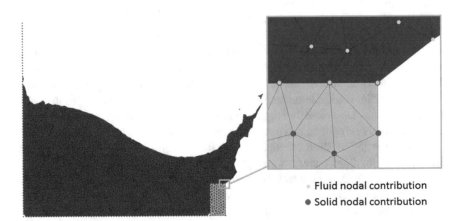

Fig. 4.1 Graphic representation of domain contributions to the global linear system

In order to ensure the coupling, the fluid and the solid meshes must have in common the nodes along the interface. In other words, there must be a node to node conformity. In this work, the solid is solved using the FEM while the fluid is computed by the PFEM. In terms of meshing, this means that the solid domain maintains the same grid during all the analysis, whereas the fluid is remeshed whenever its discretization becomes excessively distorted. The conformity of the fluid and solid meshes on the interface is guaranteed by exploiting the capability of the PFEM for detecting the boundaries. In practice, the fluid detects the solid interface nodes (nodes that are located on the external boundaries of the solid fixed mesh) in the same way it recognizes its rigid contours. As it has been explained in Sect. 3.4.1.1, this is done by an efficient combination between the Alpha Shape method and the Delaunay triangulation. According to this strategy, if the separation of the fluid contour from the solid domain is small enough so that that the Alpha Shape criteria are fulfilled, a fluid element connecting the fluid domain to the solid domain is generated. Otherwise the two domains keep apart from each other. In Fig. 4.2 a graphic representation of the method for detecting the interface is given [1].

In the situation described in the pictures of Fig. 4.2a none of the contact elements generated by the Delaunay triangulation fulfill the Alpha Shape criteria. Hence for the forthcoming time step interval there is not interaction between the solid and the fluid domains. Instead in Fig. 4.2b some hybrid elements that share solid and fluid

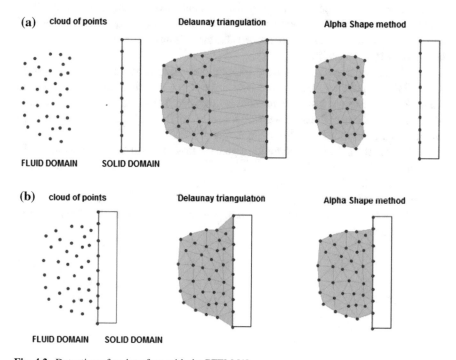

Fig. 4.2 Detection of an interface with the PFEM [1]

nodes have been generated. In this case, the coupling is active and the interface nodes assemble as described before.

For the good use of this procedure, it is essential to select a similar mesh size for the fluid and the solid domain in the interface zone. Otherwise, some typical drawbacks may arise. For example, if the solid mesh is much finer than the fluid one some distorted elements may form in the interface zone. The opposite case is even worse. If the solid mesh at the boundaries is much coarser than the fluid one, some fluid particles may cross the interface and compromise the whole computation. Using the same mesh size (and a reasonable time step increment) neither of these problems arise. Note that this behavior is characteristic of PFEM. In fact all this can occur also in the contact zone of a fluid domain with a rigid boundary in a fluid dynamic analysis, as explained in the Sect. 3.4.1. This confirms again that, from the remeshing point of view, there is not difference between the contour nodes of a deformable body and the nodes forming a rigid wall.

4.3   Coupling with the Velocity Formulation for the Solid

The coupling of the PFEM-FIC stabilized VP formulation with the V-element for the solid consists essentially on assembling adequately the linear momentum equations. For each degree of freedom for the interface nodal velocity, the solid and the fluid contributions sum. On the other hand, the stabilized continuity equation is assembled and solved only for the fluid elements because the pressure is not an unknown variable of the Velocity formulation.

In order to avoid ambiguities in the notation, the variables and the matrices referred to the fluid domain will be marked by subscript 'f', and those ones belonging to the solid by 's'. When a shared variable is considered (for example, the unknowns referred to the nodes of the interface) the subscript will be 's, f'. Instead, the subscript that marks all the nodes of the domain is '$s + f$'.

In Box 12, the solution scheme for a generic time interval $[n, n + 1]$ is given. In the scheme only the contributions of the interface nodes are analyzed. For the rest of nodes the problem is locally uncoupled, hence the schemes described in Box 3 for hypoelastic solids and in Box 8 for Newtonian fluids still hold.

All the matrices and the vectors that appear in Box 12 were already defined in Boxes 4 and 9.

For each iteration i:

1. Compute the nodal velocity increments $\Delta \bar{v}_{s,f}$:

$$K^i_{s,f} \Delta \bar{v}_{s,f} = R^i_{s,f}(^{n+1}\bar{v}^i_{s,f}, {}^{n+1}\bar{p}^i_f)$$

where:

$$K^i_f = K^\mu(^{n+1}\bar{x}^i, c^\mu) + K^\kappa(^{n+1}\bar{x}^i, c^\kappa) + K^g(^{n+1}\bar{x}^i, \sigma^i_f) + K^\rho(^{n+1}\bar{x}^i)$$

and

$$K^i_s = K^m(^{n+1}\bar{x}^i, c^{\sigma J}) + K^g(^{n+1}\bar{x}^i, \sigma^i_s) + K^\rho(^{n+1}\bar{x}^i)$$

2. Update the nodal velocities: $\quad {}^{n+1}\bar{v}^{i+1}_{s,f} = {}^{n+1}\bar{v}^i_{s,f} + \Delta \bar{v}_{s,f}$

3. Update the nodal coordinates: $\quad {}^{n+1}\bar{x}^{i+1}_{s,f} = {}^{n+1}\bar{x}^i_{s,f} + \bar{u}_{s,f}(\Delta \bar{v}_{s,f})$

4. Compute the fluid nodal pressures $\bar{p}^{i+1}_f$:

$$H_f {}^{n+1}\bar{p}^{i+1}_f = F_{p,f}(^{n+1}\bar{v}^{i+1}_f, {}^{n+1}\bar{p}^i_f)$$

where: $H_f = \left(\frac{1}{\Delta t}M_1 + \frac{1}{\Delta t^2}M_2 + L + M_b\right)$ and $F_{p,f} = Q^T {}^{n+1}\bar{v}^{i+1}_f + f_{p,i}$

5. Compute the updated stress measures:

$$^{n+1}\sigma'^{i+1}_f = 2\mu d'_f(^{n+1}\bar{v}^{i+1}_f) \quad ; \quad {}^{n+1}\sigma^{i+1}_f = {}^{n+1}\sigma'^{i+1}_f + {}^{n+1}p^{i+1}_f I$$

$$^{n+1}\sigma^{\nabla,i+1}_s = c^{\sigma J} : d_s\left(\bar{v}^{i+1}_s\right) \;\rightarrow\; {}^{n+1}\sigma^{i+1}_s = {}^n\hat{\sigma}_s + \Delta t\, {}^{n+1}\sigma^{\nabla,i+1}_s$$

6. Check the convergence: $\quad \| {}^{n+1}R^{i+1}_{s+f}(^{n+1}\bar{v}^{i+1}_{s,f}, {}^{n+1}\bar{p}^{i+1}_f) \| < tolerance$

If condition 6 is not fulfilled, return to 1 with $i \leftarrow i+1$.

At the end of each time step, for the solid elements compute

$$^{n+1}\hat{\sigma}_s = {}^{n+1}\sigma_s + \Delta t\Omega_s(^{n+1}\bar{v}_s, {}^{n+1}\sigma_s)$$

Box 12. Iterative solution scheme for FSI problem solved with the V-element for the solid

4.4   Coupling with the Mixed Velocity–Pressure Formulation for the Solid

The coupling of the fluid stabilized VP formulation with the VP-element is performed similarly to the V-element. In particular, the solution scheme of the linear momentum equations does not change, because also in this case the degrees of freedom of the velocity are not duplicated. Hence the fluid velocity coincides with the solid velocity.

The main differences concern the pressure solution. The solid and the fluid pressures are two different degrees of freedom. As a consequence, the continuity equations are solved separately for the fluid and the solid. For the solid, both the stabilized and the not stabilized equations, Eqs. (2.71) and (2.97) respectively, can be solved. Depending on the compressibility of the solid bodies one may choose one of the two schemes.

For the fluid counterpart, matrix K^κ (Eq. (3.61)) is computed using the pseudo bulk modulus κ_p and its value is predicted using the strategy described in Sect. 3.4.3.2.

In Box 13, the solution scheme for a FSI problem solved using the VP-element or the VPS-element for the solid is given for a time interval $[n, n+1]$. Once again, all the matrices and vectors that appear Box 13 refer only to the contributions of the interface nodes and they have already been defined in Box 6 for the VP-element, Box 11 for the VPS-element and Box 9 for the stabilized VP formulation for quasi-incompressible Newtonian fluids.

For the nodes that do not belong to the interface, the schemes described in Box 5 and 8 still hold.

For each iteration i:

1. Compute the nodal velocity increments $\Delta \bar{v}_{s,f}$:

$$K^i_{s,f} \Delta \bar{v}_{s,f} = R^i_{s,f}(^{n+1}\bar{v}^i_{s,f}, {}^{n+1}\bar{p}^i_{s,f})$$

where:
$$K^i_f = K^\mu(^{n+1}\bar{x}^i, c^\mu) + K^\kappa(^{n+1}\bar{x}^i, c^\kappa) + K^g(^{n+1}\bar{x}^i, \sigma^i_f) + K^\rho(^{n+1}\bar{x}^i)$$
and
$$K^i_s = K^m(^{n+1}\bar{x}^i, c^\sigma) + K^g(^{n+1}\bar{x}^i, \sigma^i_s) + K^\rho(^{n+1}\bar{x}^i)$$

2. Update the nodal velocities: $^{n+1}\bar{v}^{i+1}_{s,f} = {}^{n+1}\bar{v}^i_{s,f} + \Delta \bar{v}_{s,f}$

3. Update the nodal coordinates: $^{n+1}\bar{x}^{i+1}_{s,f} = {}^{n+1}\bar{x}^i_{s,f} + \bar{u}_{s,f}(\Delta \bar{v}_{s,f})$

---

4. Compute the fluid nodal pressures $\bar{p}_f^{i+1}$:

$$H_f{}^{n+1}\bar{p}_f^{i+1} = F_{p,f}(^{n+1}\bar{v}_f^{i+1}, {}^{n+1}\bar{p}_f^i)$$

where: $H_f = \left(\frac{1}{\Delta t}M_1 + \frac{1}{\Delta t^2}M_2 + L + M_b\right)$ and $F_{p,f} = Q^{T}\,^{n+1}\bar{v}_f^{i+1} + f_{p,i}$

5. Compute the solid nodal pressures $\bar{p}_s^{i+1}$:

$$H_s{}^{n+1}\bar{p}_s^{i+1} = F_{p,s}(^{n+1}\bar{v}_s^{i+1}, {}^{n+1}\bar{p}_s^i)$$

where for the VP-element: $H_s = \frac{1}{\Delta t}M_{1s}$ and $F_{p,s} = Q^{T}\,^{n+1}\bar{v}_s^{i+1} + g$

and for the VPS-element:

$$H_s = \left(\frac{1}{\Delta t}M_{1s} + \frac{1}{\Delta t^2}M_{2s} + L + M_b\right) \text{ and}$$

$$F_{p,s} = \frac{M_{1s}}{\Delta t}{}^n\bar{p}_s + \frac{M_{2s}}{\Delta t^2}\left({}^n\bar{p}_s + {}^n\dot{\bar{p}}_s\Delta t\right) + Q^{T}\,^{n+1}\bar{v}_s^{i+1} + f_{ps,i}$$

6. Compute the updated stress measures:

$$^{n+1}\sigma_f'^{i+1} = 2\mu d'_f(^{n+1}\bar{v}_f^{i+1}) \quad ; \quad ^{n+1}\sigma_f^{i+1} = {}^{n+1}\sigma_f'^{i+1} + {}^{n+1}p_f^{i+1}I$$

$$^{n+1}\sigma_s^{i+1} = {}^n\hat{\sigma}_s + {}^{n+1}\bar{p}_s^{i+1}I + 2\mu\Delta t\left[I' : d_s\left(\bar{v}_s^{i+1}\right)\right]$$

7. Check the convergence: $\| {}^{n+1}R_{s+f}^{i+1}(^{n+1}\bar{v}_{s,f}^{i+1}, {}^{n+1}\bar{p}_{s,f}^{i+1}) \| < tolerance$

If condition 7 is not fulfilled, return to 1 with $i \leftarrow i+1$.

At the end of each time step, for the solid elements

$$^{n+1}\hat{\sigma}_s = {}^{n+1}\sigma_s + \Delta t\Omega_s\,(^{n+1}\bar{v}_s, {}^{n+1}\sigma_s)$$

Box 13. Iterative solution scheme for FSI problem solved with the VP-element or/and the VPS-element for the solid

4.5   Numerical Examples

Falling of a Cylinder in a Viscous Fluid

This example is the two-dimensional abstraction of the moving of a circular cylinder between two parallel walls. The cylinder moves perpendicularly to its axis due to the gravity force increasing the falling velocity until an asymptotic value.

The distance from the rigid walls and the axis of the cylinder is $l = 0.02$ m. The radius of the circle is $a = 0.0025$ m. The geometry of the problem and the material data are given in Fig. 4.3 and in Table 4.1.

The same problem has been studied in many other works [3, 4]. The numerical solution can be also compared to the analytical study of the motion of a rigid cylinder with constant velocity U between two parallel plane walls. For this problem, the

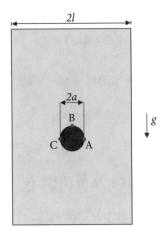

Fig. 4.3 Falling of a cylinder in a viscous fluid. Initial geometry [2]

Table 4.1 Falling of a cylinder in a viscous fluid. Problem data

Geometric data	
l	0.02 m
a	0.0025 m
g	9.81 m/s^2
Fluid data	
Density	$1.0 \cdot 10^3$ kg/m^3
Viscosity	0.1 Pa $\cdot$ s
Solid data	
Density	$1.2 \cdot 10^3$ kg/m^3
Young modulus	10^7 GPa
Poisson ratio	0.35

analytical solution is obtained studying the creeping motion equations using stream functions [5]. Considering stick conditions on the walls, for a unit of length of the cylinder the drag force in stationary conditions is

$$F = \frac{4\pi\mu U}{ln(l/a) - 0.9157 + 1.7244(a/l)^2 - 1.7302(a/l)^4} \tag{4.1}$$

where U is the velocity of the rigid cylinder.

This relation for the drag force holds also for the stationary conditions of the problem studied in this section. From the equilibrium of the drag force with the Archimedes' force yields

$$F = \Delta\rho g \pi a^2 \tag{4.2}$$

where $\Delta\rho$ is the difference between the density of the two materials involved. Combining Eqs. (4.1) and (4.2), the terminal velocity of a rigid circular cylinder with radius a is

$$U_{max} = \frac{ln(l/a) - 0.9157 + 1.7244(a/l)^2 - 1.7302(a/l)^4}{4\mu}\Delta\rho g a^2 \tag{4.3}$$

For this problem this relation gives

$$U_{max} = 0.0365 \text{m/s} \tag{4.4}$$

This value for the terminal velocity is near to the asymptotic ones obtained by the mentioned works [3, 4].

Numerical studies show that this expression holds only for a tight range of values of the viscosity and, specifically, is not suitable for a fluid with small viscosity.

In this work, the solid has been modeled not as a rigid body but with a hypoelastic model, and using a high value for the Young modulus. The VP formulation has been used for the solid. This means that the example belongs properly to the class of the FSI problems. The problem has been solved for both stick and slip conditions on the vertical walls of the container. Due to the dominant role of the viscosity, the solution of the stick problem does not tend to the solution of the slip problem.

In Figs. 4.4 and 4.5 some snapshots of the simulation with the velocity contours are given. Figure 4.4 refers to the slip case, Fig. 4.5 to the stick one.

The numerical results plotted in Figs. 4.4 and 4.5 have been obtained with a mesh with average size 0.0007 m. The pictures show the importance of the boundary conditions for this case. It is evident that the velocity fields obtained for the slip and the stick cases, as well the terminal velocity of the cylinder, are significantly different.

For the same mesh, the resulting pressure field for the slip and stick cases is illustrated in Fig. 4.6. The pictures show that the perturbation caused by the motion of the cylinder is almost imperceptible and there are not significant differences between the slip and stick cases.

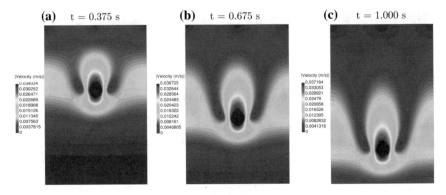

Fig. 4.4 Falling of a cylinder in a viscous fluid. Snapshots of the cylinder motion at different instants of the 2D simulation (slip conditions on the boundaries). Velocity contours are depicted over the solid and fluid domains

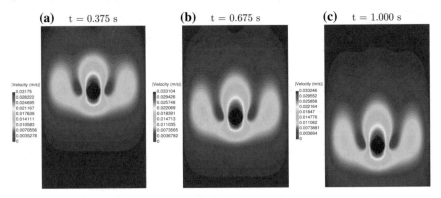

Fig. 4.5 Falling of a cylinder in a viscous fluid. Snapshots of the cylinder motion at different instants of the 2D simulation (stick conditions on the boundaries). Velocity contours are depicted over the solid and fluid domains [2]

In the graph of Fig. 4.7 the time evolution of the vertical velocity of the cylinder obtained with the finest meshes (average size = 0.0004 m) are given for both the slip and stick cases.

The terminal velocities of the cylinder obtained with slip and stick conditions are 0.0377 and 0.0336 m/s, respectively.

For this example the transmission conditions between the solid and the fluid domain have been monitored. The curves of Fig. 4.8 represent the time evolution of the Neumann condition in the X-direction (horizontal) at the points A, B, C located at the boundary of the cylinder and depicted in Fig. 4.3. Specifically, the value plotted in the curves is the mean value of the X-component of vector $\boldsymbol{\sigma n}$ $(\sigma_{xx}n_x + \tau_{xy}n_y)$ computed for the fluid and the solid elements at the points A, B, C of Fig. 4.3. The graph shows that the transmission condition is guaranteed during all the analysis.

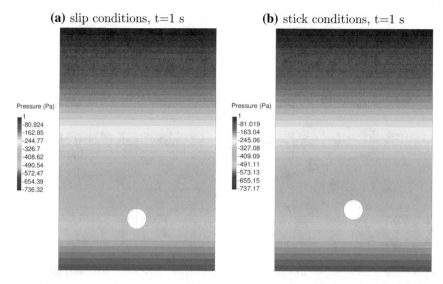

(a) slip conditions, t=1 s   **(b)** stick conditions, t=1 s

Pressure (Pa)
1
-80.924
-162.85
-244.77
-326.7
-408.62
-490.54
-572.47
-654.39
-736.32

Pressure (Pa)
1
-81.019
-163.04
-245.06
-327.08
-409.09
-491.11
-573.13
-655.15
-737.17

Fig. 4.6 Falling of a cylinder in a viscous fluid. Pressure field obtained for the slip and stick cases [2]

Fig. 4.7 Falling of a cylinder in a viscous fluid. Time evolution of the vertical velocity of the cylinder. Results for the slip and the stick cases [2]

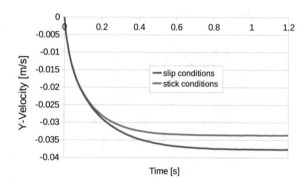

A convergence study was also performed. In Fig. 4.9 the coarsest (average mesh size − 0.0015 m) and the finest (average mesh size = 0.0004 m) meshes used for the example are depicted on the solid body. The whole domain has been discretized using 2983 and 42,873 3-noded triangular elements, respectively.

The graph of Fig. 4.10 shows the time evolution of the vertical velocity obtained with five of the meshes tested for the slip case.

In Fig. 4.11 the results obtained with the same meshes assuming stick conditions on the walls are given.

The convergence graphs for both the stick and slip conditions are plotted in Fig. 4.12. Both cases show the same convergence rate.

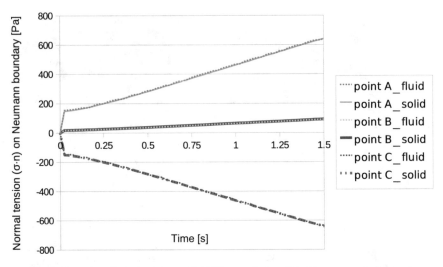

Fig. 4.8 Falling of a cylinder in a viscous fluid. Time evolution of the X-component of $\sigma_{xx}n_x + \sigma_{xy}n_y$ computed at the point A, B, C of Fig. 4.3 [2]

(a) Mesh size= 0.0015m   **(b)** Mesh size= 0.0004m

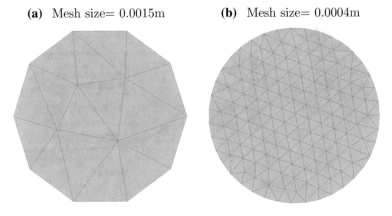

Fig. 4.9 Falling of a cylinder in a viscous fluid. Coarsest (average mesh size $= 0.0015$) and finest (average mesh size $= 0.0004$) meshes used for the example depicted on the solid domain

The problem has been solved also for a quasi-incompressible solid using the VPS-element. For this case, a Poisson ratio of 0.4999 and the same Young modulus used for the problem solved with the VP-element, have been considered. In Fig. 4.13 the velocity and the pressure fields for the solid and the fluid computed at $t = 1$ s for stick conditions on the walls and using a mean mesh size of 0.007 m are given.

In the graph of Fig. 4.14 the time evolution of the vertical velocity obtained with the VPS-element for $\nu = 0.4999$ is compared with the solution obtained with the VP-element for $\nu = 0.35$ and the same average mesh size and boundary conditions.

Fig. 4.10 Falling of a cylinder in a viscous fluid. Time evolution of the vertical velocity of the cylinder for five different meshes and slip conditions on the walls

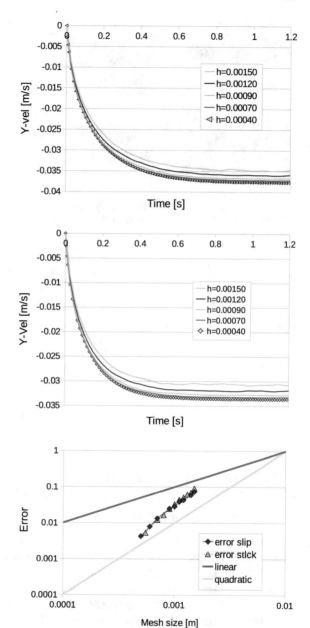

Fig. 4.11 Falling of a cylinder in a viscous fluid. Time evolution of the vertical velocity of the cylinder for five different meshes and stick conditions on the walls

Fig. 4.12 Falling of a cylinder in a viscous fluid. Convergence curves for the slip and stick cases

(a) Velocity field   **(b)** Fluid pressure   **(c)** Solid Cauchy stress

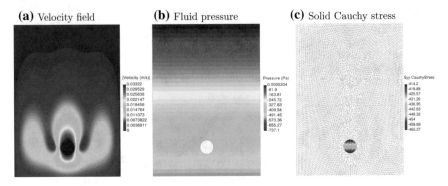

Fig. 4.13 Falling of a quasi incompressible cylinder in a viscous fluid. Velocity field, fluid pressure and solid Cauchy stress (YY component) at $t = 1$ s for stick conditions on the walls and a mean mesh size of 0.007 m [2]

Fig. 4.14 Falling of a cylinder in a viscous fluid. Solutions obtained using for the solid the VP-element ($\nu = 0.35$) and the VPS-element ($\nu = 0.4999$) [2]

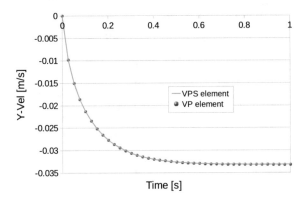

As expected, the solutions are almost the same and the vicinity to the incompressible limit does not compromise the quality of the results.

Water Entry of a Nylon Sphere

The problem was presented by Aristoff et al. in [6]. In the mentioned work the experimental results of the water entry of spheres of different materials are studied.

In this section, the case of a nylon sphere is analyzed. The numerical results given by the Unified formulation (with the VP-element for the solid) are compared to the results of the laboratory test. The sphere impacts the water in the tank with a vertical velocity of 2.17 m/s and its diameter is 2.54 cm. The density of nylon is 1140 kg/m^3 and the Young modulus and Poisson ratio are 3 GPa and 0.2, respectively. The water was modeled considering a density of 1000 kg/m^3, a dynamic viscosity 0.00089 Pa · s and a bulk modulus 2.15 GPa. In order to simulate this problem correctly, a very fine mesh is necessary. For this reason the whole domain was discretized with 1059924 tetrahedra. The time step used for the analysis is $\Delta t = 10^{-4}$ s.

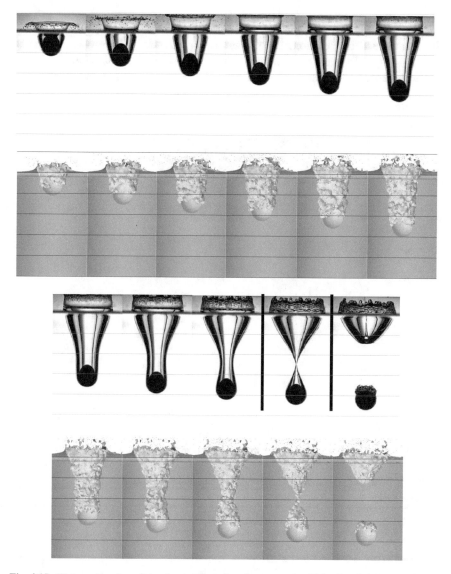

Fig. 4.15 Water entry of a nylon sphere: comparison between experimental and numerical results

In Fig. 4.15 the numerical results are compared to the experimental ones [6]. The comparison shows the good agreement of the results obtained with the PFEM Unified formulation with the laboratory values and its capability to simulate the complex phenomena such as the creation of the void (the air has not been taken into account) within the fluid domain.

Filling of an Elastic Container with a Viscous Fluid

This example is inspired from a similar problem presented in [1]. A volume of a viscous fluid drops from a rigid container over a thin and highly deformable membrane. The impact of the fluid mass causes an initial huge stretching of the structure and its

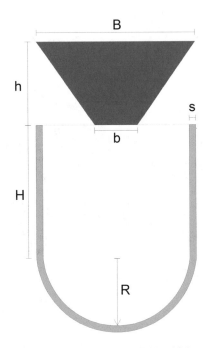

Fig. 4.16 Filling of an elastic container with a viscous fluid. Initial geometry [2]

Table 4.2 Filling of an elastic container with a viscous fluid. Problem data

Geometry data	
h	2.5
H	3.75
R	2.25
b	1.3
B	4.8714
s	0.2
Fluid data	
Viscosity	50, 100 Pa · s
Density	1000 kg/m^3
Solid data	
Young modulus	2.1 10^7 Pa
Poisson ratio	0.3
Density	20 kg/m^3

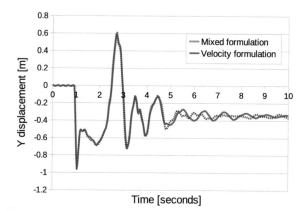

Fig. 4.17 Filling of an elastic container with a viscous fluid ($\mu = 50\,\text{Pa}\cdot\text{s}$). Vertical displacement of the bottom of the elastic container obtained using the V and the VP elements for the solid domain [2]

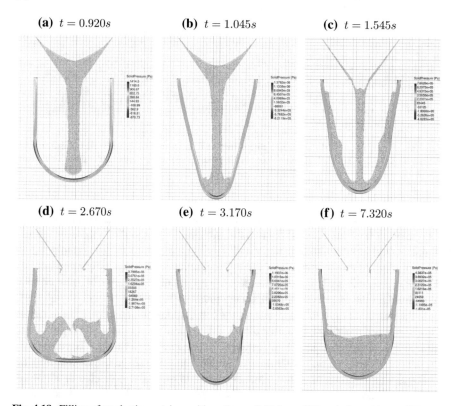

Fig. 4.18 Filling of an elastic container with a viscous fluid ($\mu = 50\,\text{Pa}\cdot\text{s}$). Snapshots at different instants of the 2D simulation. Pressure contours depicted over the solid domain [2]

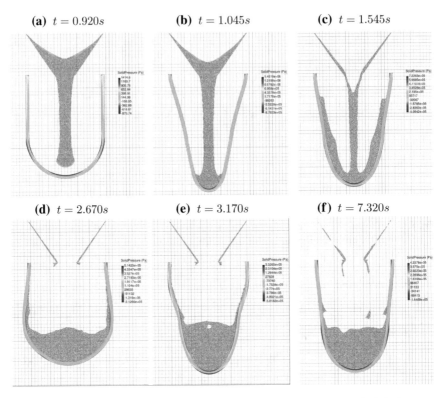

Fig. 4.19 Filling of an elastic container with a viscous fluid ($\mu = 100\,\text{Pa} \cdot \text{s}$). Vertical displacement of the bottom of the elastic container obtained using the V and the VP elements for the solid domain [2]

subsequent oscillations. A hypoelastic law is used for the material of the structure. The problem was solved in 2D for two different values of the fluid viscosity, namely 50 and 100 Pa · s, and with both the V and VP elements. The purpose was to compare the formulations and to show that both solid elements can be used for the modeling of standard elastic solids in FSI problems. The initial geometry of the problem is given in Fig. 4.16 and the material data are given in Table 4.2.

In the graph of Fig. 4.17 the results for the less viscous case ($\mu = 50\,\text{Pa·s}$) obtained using the V and the VP elements for the solid are given. The comparison is performed for the vertical displacement of the lowest point of the elastic structure. The curves are almost coincident and only after 4.5 s of simulation some slight differences appear.

For the same problem, some representative snapshots are collected in Fig. 4.18. The pressure contours are depicted over the solid domain, while over the fluid one the mesh is plotted. The numerical results refer to the simulation using the VP-element for the solid.

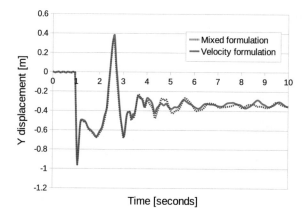

Fig. 4.20 Filling of an elastic container with a viscous fluid ($\mu = 100\,\text{Pa} \cdot \text{s}$). Snapshots at different instants of the 2D simulation. Pressure contours depicted over the solid domain [2]

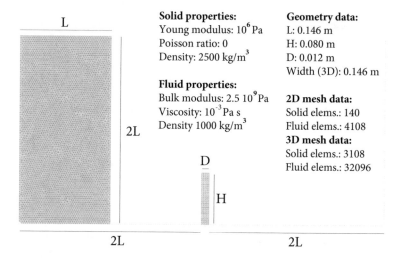

Solid properties:
Young modulus: 10^6 Pa
Poisson ratio: 0
Density: 2500 kg/m^3

Fluid properties:
Bulk modulus: 2.5 10^9 Pa
Viscosity: 10^{-3} Pa s
Density 1000 kg/m^3

Geometry data:
L: 0.146 m
H: 0.080 m
D: 0.012 m
Width (3D): 0.146 m

2D mesh data:
Solid elems.: 140
Fluid elems.: 4108
3D mesh data:
Solid elems.: 3108
Fluid elems.: 32096

Fig. 4.21 Collapse of a water column on a deformable membrane. Initial geometry and problem data

In Fig. 4.20 the snapshots of the most viscous case ($\mu = 100\,\text{Pa} \cdot \text{s}$) are given for the same time instants of Fig. 4.18. The numerical results refer to the solution obtained using the VP-element for the solid domain.

Also for the most viscous case, the results obtained using the velocity and the mixed Velocity–Pressure formulations for the solid are compared analyzing the time evolution of the vertical displacement at the bottom of the elastic structure. The two curves of Fig. 4.19 represent the solutions obtained with the velocity and the mixed Velocity–Pressure formulations.

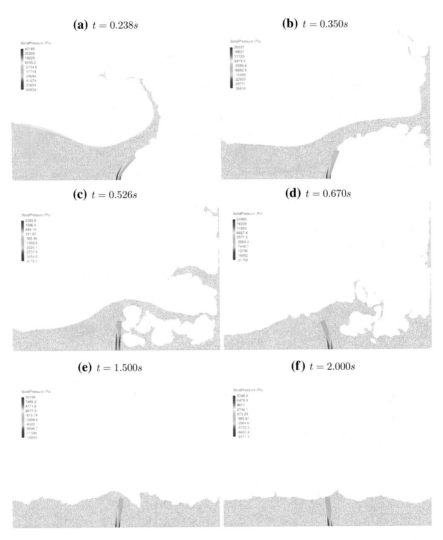

Fig. 4.22 Collapse of a water column on a deformable membrane. Snapshots of the 2D simulation at different instants. The VP element is used for the solid [2]

Once again, the differences between the results of the two formulations are very small and this is a further evidence of the validity and flexibility of the proposed Unified formulation. This example evidenced the possibility to choose for the solid a velocity or a mixed formulation also for solving FSI problems.

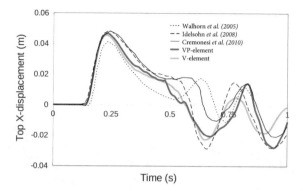

Fig. 4.23 Collapse of a water column on a deformable membrane. Snapshots of the 3D simulation of different instants. The V-element is used for the solid

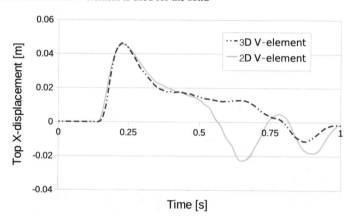

Fig. 4.24 Collapse of a water column on a deformable membrane. Horizontal deflection of the left top corner on time. Numerical results obtained with V and VP elements for the solid. Comparison with numerical results obtained in [1, 7, 9]

Collapse of a Water Column on a Deformable Membrane

The problem illustrated in Fig. 4.21 was introduced by Walhorn et al. [7].

The water column collapses by instantaneously removing the vertical wall. This originates the flow of water within the tank, the formation of a jet after the water stream hits the rigid ground, and the subsequent sloshing of the fluid as it impacts a highly deformable elastic membrane. The membrane bends and starts oscillating under the effect of its inertial forces and the impact with the water stream.

The problem has been solved in 2D as in 3D, considering stick conditions for the rigid walls.

In Fig. 4.22 some representative snapshots of the 2D simulation are given.

The results obtained with the present formulation have been compared to the ones computed in [1, 8, 9]. In the graph of Fig. 4.24 the time evolution of the horizontal deflection of the left top corner is illustrated.

(a) $t = 0.23s$ **(b)** $t = 0.37s$

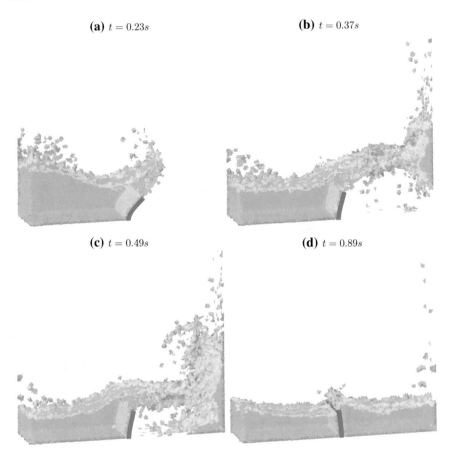

(c) $t = 0.49s$ **(d)** $t = 0.89s$

Fig. 4.25 Collapse of a water column on a deformable membrane. Horizontal deflection of the left top corner on time. Comparison between 2D and 3D analyses (V-element for the solid part)

The diagram shows that, for the first part of the analysis, the proposed formulation agrees well with the results reported in the literature. After around 0.5 s of simulation, the numerical results of each formulation starts to diverge, although for all the formulations the membrane oscillates two times around its vertical position before the time instant $t = 1$ s. The first part of the simulation is easier to analyze than the second one because the phenomena to model are less aleatory and the fluid splashes do not affect the results, as it occurs in the second part of the simulation. Furthermore, the initial deformation of the elastic structure affects highly the rest of the simulation. In fact, a smaller bending of the membrane induces an impact of the water stream at a higher height of the containing wall and with a bigger tangential component of the impact velocity. Consequently, the fluid stream impacts against the right side of the elastic membrane later and with reduced inertial forces. These considerations have an even greater effect for the 3D simulation.

In the graph of Fig. 4.25 the results of the 2D and 3D simulations obtained using the V-element are compared.

The graph shows that there is a good agreement between the 2D and 3D analyses in the first part of the graph. Then the graphs are quite different. In particular, in the 3D problem the fluid stream impacts later against the right side of the membrane producing a reduced displacement of the elastic structure. This occurs because, in the 3D simulation, the fluid stream after the impact against the vertical rigid wall (this occurs after around 0.35 s of analysis), can move in the direction transversal to the main direction of its motion and this reduces the impact force of the water stream. This behavior is intensified if stick boundary conditions are considered and the mesh is not sufficiently fine.

In Fig. 4.23 the numerical results of the 3D simulation are given.

4.6 Summary and Conclusions

In this chapter the Unified Stabilized formulation has been used for the solution of FSI problems. Specifically, the numerical schemes for coupling the stabilized Velocity–Pressure formulation for quasi-incompressible Newtonian fluids with the V, the VP and the VPS elements for the solid have been presented. It has been shown that, depending on the specific need, one may choose any of these solid elements.

It has been shown that the implementation effort for extending the Unified formulation to the coupled FSI problems is small. Essentially, it consists of assembling adequately the global linear system and exploiting the capability of the PFEM for the detection of the interface.

Several validating examples have been given. The numerical solution obtained with the proposed scheme has been compared to analytical solutions and numerical results of other formulations for free surface FSI problems. Good agreement has been found in all cases. It has also been shown that the method is convergent.

References

1. M. Cremonesi. *Doctoral thesis: A Lagrangian Finite Element Method for the Interaction Between Flexible Structures and Free Surfaces Fluid Flows.* 2010.
2. A. Franci, E. Oñate, and J. M. Carbonell. Unified lagrangian formulation for solid and fluid mechanics and fsi problems. *Computer Methods in Applied Mechanics and Engineering,* 298:520–547, 2016.
3. A.J. Gil, A. Arranz Carreño, J. Bonet, and O.Hassan. The immersed structural potential method for haemodynamic applications. *Journal of Computational Physics,* 229:8613–8641, 2010.
4. X. Wang and W.K. Liu. Extended immersed boundary method using fem and rkpm. *Computer Methods in Applied Mechanics and Engineering,* 193:1305–1321, 2004.
5. J. Happel and H. Brenner. Low reynolds number hydrodynamics.
6. J. Aristoff, T.T. Truscott, A.H. Techet, and J.W.M Bush. The water entry of decelerating spheres. *Physics of Fluids,* 22:032102, 2010.

7. E. Walhorn, A. Kolke, B. Hubner, and D.Dinkler. Fluid-structure coupling within a monolithic model involving free surface flows. *Computer & Structures Methods in Applied Mechanics and Engineering*, 83 (25–26):2100–2111, 2005.
8. B. Hubner, E. Walhorn, and D.Dinkler. A monolithic approach to fluid-structure interaction using space-time finite elements. *Computer Methods in Applied Mechanics and Engineering*, 193:2087–2104, 2004.
9. S.R. Idelsohn, J. Marti, A. Limache, and E. Oñate. Unified lagrangian formulation for elastic solids and incompressible fluids: Applications to fluid-structure interaction problems via the pfem. *Computer Methods In Applied Mechanics And Engineering*, 197:1762–1776, 2008.

Chapter 5
Coupled Thermal–Mechanical Formulation

5.1 Introduction

This chapter is devoted to the coupling of the Unified formulation with the heat transfer problem. The objective is to solve general problems involving fluids, solids or both of them which behavior is temperature-dependent and where the phase change of the materials is allowed.

There is a growing interest in industry in having a technology capable to solve coupled thermo-mechanical problems. In fact, industrial problems generally lack of an analytical solution and often the unique way to predict their solution is based on experience. An alternative is represented by experimental tests. However, these cannot be carried out for all the required problems due to the complexity of the geometry or to the high associated cost. For these reasons, computer-based methods represent often the most reliable alternative for the analysis of coupled thermal–mechanical problems.

Nevertheless, thermo-mechanical problems involving fluid and solids represent a challenge also for the numerical analysis due to their multi-field nature and high non-linearity. From the computational point of view, there are many complications associated to these problems.

First of all, the coupling with the heat problem increases the non-linearity of the mechanical problem. The solution of the mechanical problem depends on the heat transfer via temperature dependence of the material properties, at the same time, the heat problem depends on the solution of the mechanical problem due to the change of the configuration.

The complexity of the problem increases if phase change is also considered. This phenomenon not only increases the non-linearity of the problem, but it also requires the implementation of a specific technology for its simulation. Furthermore accounting for phase change may induce problems related to the quality of the mesh. Consider, for example, the melting of a solid body. If a Lagrangian mesh is used, as it generally occurs with solids, the large deformations undergone by the melting body may compromise the quality of the mesh and the results of the overall simulation.

© Springer International Publishing AG 2017
A. Franci, *Unified Lagrangian Formulation for Fluid and Solid Mechanics,*
Fluid-Structure Interaction and Coupled Thermal Problems Using the PFEM,
Springer Theses, DOI 10.1007/978-3-319-45662-1_5

Another critical point is the imposition of the boundary conditions for the temperature. For this reason it is required to track the contours of the domains involved in the simulation for all the duration of the analysis. This task may not be trivial if the problem involves complex geometries, severe changes of topology or phase change.

With the PFEM many of these complexities are overcome thanks to its Lagrangian nature. In fact, as it has already explained in the previous chapters, the boundaries of the bodies are automatically detected by the position of the boundary nodes. Furthermore, if PFEM is also used for modeling the solid dynamics, the drawbacks associated to the large deformations of the solid caused by heat are overcome through the remeshing which guarantees a fine discretization for all the duration of the analysis. Finally, in the Lagrangian formulation the convective term disappears from the heat transfer equation. For all these reasons, other authors faced this problem in the past using the PFEM [1–3].

In this work, the PFEM is only used for the fluid domain. So in order to overcome the decay of quality of the solid mesh in a melting process, a further technology is required. In this work, solid elements transform into fluid elements when they fulfil a phase-change criterion. If this occurs, the melted solid element is considered and computed as a fluid. Consequently it benefits from the PFEM technology and it can be remeshed if required. The details of the algorithm will be given in the following sections.

Due to the complexity and the vastness of the issue, in this treatise various assumptions and simplifications have been accepted. This has been done in some cases at the expense of the accuracy of the analyses. Nevertheless, the main objective of this chapter is to show that the PFEM Unified formulation for FSI problems can be easily coupled to the heat problem in order to solve simulations where also the temperature affects. The principal assumptions of this work are described in the following part.

In all the simulations, the air has not been taken into account, so heat transmission only occurs via the contact between solid and liquid bodies, or through the contours where the boundary conditions for the temperature have been imposed. Furthermore, in the phase change modeling, there is not a transition where a multi-fluid analysis is employed. When an element fulfils the phase change criteria, it takes instantaneously the physical properties of the other material involved in the analysis, without considering a layer formed by a transition material. Note that these simplifications have been assumed just for convenience and they are not required by the formulation itself. With an additional implementation work both assumptions can be avoided. A proof of this are the successful attempts made in previous works where the PFEM was successfully used for modeling multi-fluids [4].

In order to guarantee a strong thermal–mechanical coupling, the heat problem is solved within the same iteration loop used for solving the mechanical problem. This strategy increases the non-linearity of the mechanical problem because of its dependence on the temperature. A simpler way to solve the global problem is to consider a constant temperature during the non-linear iterations of the mechanical problem and to solve the heat problem only once the mechanical problem has converged. This approach leads to a lower computational cost and it reduces the non-linearity of the mechanical problem. However, the resulting coupling is weaker. Both strong and weak coupling strategies belong to the staggered solution schemes because the mechanical and the heat problem are solved in different linear systems. Conversely, in monolithic approaches both problems are solved simultaneously. On the one hand, this scheme enables the stability and the convergence of the whole coupled problem, on the other, it leads to longer and worse conditioned algebraic systems. However, in this work this scheme has not been chosen essentially for another reason. Monolithic schemes require modifications also for the mechanics solution, because in the assembly of the linear system also the degrees of freedom of the temperature have to be included. Instead, in staggered schemes, the FSI formulation remains the same because the mechanical and the heat transfer problems are two different blocks and they may be implemented separately. This, apart from reducing the implementation effort, gives us the possibility to test the Unified formulation in its original version.

This chapter is split into the following sections. First, the governing equations of the heat are introduced and discretized using FEM. Also the scheme for the solution of a general time step is given. Then the thermal coupling algorithm is described and validated with three numerical examples. After that, the strategy for modeling phase change is explained. The chapter ends with an example where the thermal–mechanical coupling technique is applied.

5.2 Heat Problem

The heat transfer problem in a Lagrangian setting is governed by the following differential equation

$$\rho c \frac{\partial T}{\partial t} - \frac{\partial}{\partial x_i}\left(k \frac{\partial T}{\partial x_i}\right) + Q = 0 \quad i = 1, n_s \quad \text{in } \Omega \tag{5.1}$$

where T is the temperature, c is the thermal capacity, k is the heat conductivity and Q is the heat source.

Notice that the convective term does not appear in Eq. (5.1).

The temperature and the heat flux at the boundaries are prescribed with the following boundary conditions

$$\phi - \phi^p = 0 \quad \text{on } \Gamma_\phi \tag{5.2}$$

$$k \frac{\partial \phi}{\partial n} + q_n^p = 0 \quad \text{on } \Gamma_q \tag{5.3}$$

where ϕ^p and q_n^p are the prescribed temperature and the prescribed normal heat flux at the boundaries Γ_ϕ and Γ_q, respectively, and n is the direction normal to the boundary.

The problem is completed by the initial condition

$$T(t = 0) = \bar{T} \quad \text{in } \Omega \tag{5.4}$$

The space for the test functions for the temperature is defined as

$$\hat{w}_i \in U_0, \qquad U_0 = \{\hat{w}_i | \hat{w}_i \in C^0, \ \hat{w}_i = 0 \ on \ \Gamma_\phi\} \tag{5.5}$$

Multiplying Eq. (5.1) by the test functions and integrating over the updated configuration domain, the following global integral form is obtained

$$\int_\Omega \hat{w}_i \left[\rho c \frac{\partial T}{\partial t} - \frac{\partial}{\partial x_i} \left(k \frac{\partial T}{\partial x_i} \right) + Q \right] d\Omega = 0 \tag{5.6}$$

Integrating Eq. (5.6) by parts, using Eq. (5.3) and neglecting the space changes of the
conductivity, the following weak variational form of the heat problem is obtained

$$\int_\Omega \hat{w} \rho c \frac{\partial T}{\partial t} d\Omega + \int_\Omega \frac{\partial \hat{w}}{\partial x_i} k \frac{\partial T}{\partial x_i} d\Omega + \int_\Omega \hat{w} Q d\Omega + \int_{\Gamma_q} \hat{w} q_n^p d\Gamma = 0 \tag{5.7}$$

5.2.1   FEM Discretization and Solution for a Time Step

For the heat problem, the same procedure for the spatial discretization of the mechanical one is used. Hence the analysis domain is discretized into finite elements with n
nodes in the standard manner, leading to a mesh with N_e elements and N nodes. For
2D problems 3-noded linear triangles are used, while 3D domains are discretized
using 4-noded tetrahedra. The temperature is interpolated over the mesh in terms of
its nodal values, in the same manner as for the velocities and the pressure, using the
global linear shape functions N_j spanning over the elements sharing node j ($j = 1, N$).
In matrix form,

$$T = N_T \bar{T} \tag{5.8}$$

where $N_T = [N_1, N_2, \ldots, N_N]$ and

$$\bar{T} = \begin{Bmatrix} \bar{T}_1 \\ \bar{T}_2 \\ \vdots \\ \bar{T}_N \end{Bmatrix} \tag{5.9}$$

In Eq. (5.8) vector $\bar{T}$ contains the nodal temperatures for the whole mesh. Substituting
Eq. (5.8) into (5.7) and choosing a Galerkin formulation with $\hat{w}_i = N_i$ leads to the
following algebraic equation

$$C\dot{\bar{T}} + \hat{L}\bar{T} + f_T = 0 \tag{5.10}$$

The matrices and vectors in Eq. (5.10) are assembled from the element contributions given in Box 1.

$$C_{ij}^e = \int_{\Omega^e} \rho c N_i^e N_j^e \, d\Omega$$

$$\hat{L}_{ij}^e = \int_{\Omega^e} k(\nabla^T N_i^e)\nabla N_j^e \, d\Omega$$

$$f_{T_i}^e = \int_{\Omega^e} N_i^e Q \, d\Omega - \int_{\Gamma_q} N_i^e q_n^p \, d\Gamma$$

Box 1. Element form of the matrices and vectors in Eq. (5.10)

Equation (5.10) is solved using a standard forward Euler scheme. For the i-th iteration within a time interval $[n, n+1]$ the following linear system is solved

$$\left[\frac{1}{\Delta t}C + \hat{L}\right]\Delta\bar{T} = -{}^{n+1}\bar{r}_T^i \quad , \quad r_T = C\dot{\bar{T}} + \hat{L}\bar{T} + f_T \tag{5.11}$$

with

$$^{n+1}\bar{T}^{i+1} = {}^n\bar{T}^i + \Delta\bar{T} \tag{5.12}$$

If the following conditions are verified, the next time step is considered.

$$\|{}^{n+1}\bar{T}^{i+1} - {}^{n+1}\bar{T}^i\| \le e_T \|{}^n\bar{T}\| \tag{5.13}$$

where e_T is a prescribed error norm.

5.3 Thermal Coupling

In a thermo-coupled mechanical problem, the deformation induced by the heat has to be taken into account. The total deformation rate is the sum of the elastic, the plastic and thermal parts. Hence Eq. (2.107) for a thermal–elastoplastic problem is:

$$D = D_{el} + D_{pl} + D_{th} \tag{5.14}$$

where the thermal part of the deformation rate tensor D_{th} is defined as

$$D_{th} = \alpha\dot{T}mm^T \tag{5.15}$$

where α is the thermal expansion coefficient and $m = [1, 1, 1, 0, 0, 0]^T$.

In this work, the heat problem is solved after the mechanical problem in the same iteration loop (Fig. 5.1).

This strategy belongs to the class of staggered schemes because the heat and the mechanical problems are solved in two different linear systems. This method is called

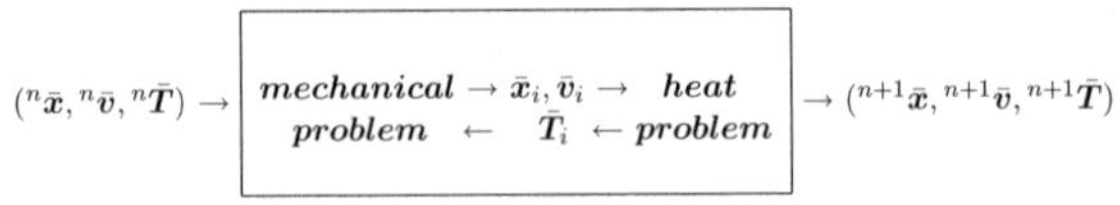

Fig. 5.1 Inner staggered
thermal coupling scheme for
a general time step $({}^{n}t, {}^{n+1}t)$

$$({}^{n}\bar{x}, {}^{n}\bar{v}, {}^{n}T) \rightarrow \boxed{\begin{array}{c} mechanical \rightarrow \bar{x}_i, \bar{v}_i \rightarrow heat \\ problem \leftarrow \quad T_i \leftarrow problem \end{array}} \rightarrow ({}^{n+1}\bar{x}, {}^{n+1}\bar{v}, {}^{n+1}T)$$

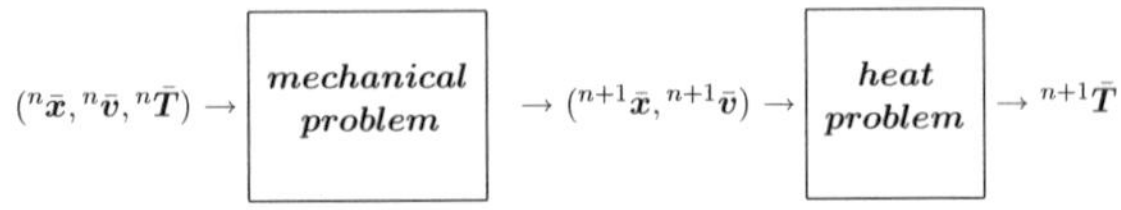

Fig. 5.2 External staggered
thermal coupling scheme for
a general time step $({}^{n}t, {}^{n+1}t)$

$$({}^{n}\bar{x}, {}^{n}\bar{v}, {}^{n}T) \rightarrow \boxed{\begin{array}{c} mechanical \\ problem \end{array}} \rightarrow ({}^{n+1}\bar{x}, {}^{n+1}\bar{v}) \rightarrow \boxed{\begin{array}{c} heat \\ problem \end{array}} \rightarrow {}^{n+1}T$$

'*internal* scheme' in order to distinguish it from another staggered scheme that will
be presented later. This strategy increases the non-linearity of the problem because
the properties of the material may depend on the temperature. The convergence is
tested at the end of each iteration for the velocities, the pressure and the temperature.

The thermal coupling can be also performed through a different staggered scheme,
that will be called '*external* scheme'. In this case the heat problem is solved after the
convergence of the mechanical problem (Fig. 5.2).

The external staggered scheme reduces the non-linearity of the mechanical problem because during the non-linear iterations the temperature is considered as a constant. It also gives the possibility to choose different time step increments for the
mechanical and the heat problems. On the other hand, the resulting coupling is weaker
than for the internal scheme. For guaranteeing a stronger coupling the mechanical
and the thermal problems should be solved iteratively within the time step. However, this strategy increases the computational cost of the analyses. In practice, their
duration increases by a factor equal to the number of the global iterations of the
thermal–mechanical problem.

5.3.1   Numerical Examples

In this section, three numerical examples are presented. First, the heat transfer problem is validated comparing the numerical solution of a heated plate to the analytical
one. Next, the sloshing of a fluid in a heated tank is presented. Finally, the heating
of three solid objects in a fluid contained in a rectangular tank is studied.

Heating of a Plate

The objective of this first example is to verify the implementation of the thermal
problem. The problem consists of a square plate at initial temperature $T = 100$ cooled
by three of its edges at constant temperature $T = 0$ keeping the other edge insulated.
Figure 5.3 shows a graphical representation of the problem.

The thermal conductivity of the plate is $k = 1$. The length of the plate edges is
$L = 1$.

The problem can be computed analytically by solving the partial differential equations of the heat problem with the proper initial and boundary conditions. The analytical solution for the temperature field is:

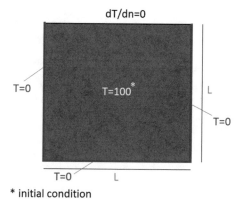

Fig. 5.3 Heating of a plate. Initial geometry, thermal boundary and initial conditions

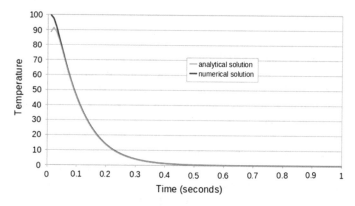

Fig. 5.4 Heating of a plate. Temperature evolution on time at the point $(x, y) = (0.5, 1)$. Analytical and numerical solutions

$$T(x, y, t) = \sum_{m=1}^{\infty} \sum_{n=1}^{\infty} \frac{1600}{(2m-1)(2n-1)\pi^2} \sin\left(\frac{(2m-1)\pi y}{2L}\right) .$$
$$\cdot \sin\left(\frac{(2n-1)\pi x}{L}\right) \exp\left(-\frac{(2m-1)^2 + 4(2n-1)^2}{4L^2}\pi^2 t\right) \quad (5.16)$$

In the graph of Fig. 5.4 the temperature evolution with time at the central point of the top edge (with coordinates $(x, y) = (0.5, 1)$) obtained with the proposed approach is compared to the analytical solution. Apart from the initial instants, the curves are almost identical.

Sloshing of a Fluid in Heated Tank

This fluid dynamics problem was presented in [5]. A fluid at initial temperature $T = 20\,°C$ oscillates due to the hydrostatic forces induced by its initial position in

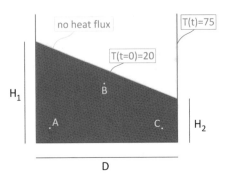

Material data
Viscosity: $10^{-2} Pa \cdot s$
Bulk modulus: $10^9 Pa$
Density: $10^3 kg/m^3$
Conductivity: $2 \cdot 10^3 W/(m \cdot {}^\circ C)$
Thermal capacity: $4 \cdot 10^3 J/(kg \cdot {}^\circ C)$
Geometry data
H_1: 0.7 m
H_2: 0.3 m
D: 1 m
Average mesh size: 0.02 m
Analysis data
Total duration: 100 s
Time step increment: 0.005 s

Fig. 5.5 2D sloshing of a fluid in a heated tank. Initial geometry, problem data, thermal boundary and initial conditions [5]

a rectangular tank heated to a uniform and constant temperature of $T = 75\,^\circ C$. The geometry and the problem data of the 2D simulation are shown in Fig. 5.5.

The fluid domain has been initially discretized with 2828 3-noded triangles. The coupled thermal-fluid dynamics simulation has been run for 100 s using a time step increment of $\Delta t = 0.005$ s.

The main purpose of this example is to show the capability of the proposed Lagrangian technique for dealing with thermal coupled problems involving severe changes of topology. For this reason, some simplifications have been accepted. For example, the properties of the fluids do not depend on temperature. Furthermore, with the purpose of reducing the computational time and visualizing better the temperature contours, a very high (and not realistic) thermal conductivity has been used. These assumptions reduce clearly the truthfulness of the problem but, as it has already pointed out, this is not the principal objective of this example.

In Fig. 5.6 the evolution of the temperature with time at the points A, B and C of Fig. 5.5 is plotted. The coordinates of these sampling points are (0.1 m, 0.1 m), (0.5 m, 0.4 m) and (0.9 m, 0.1 m), respectively.

The fluid, because of its high thermal conductivity, changes its temperature quickly due to the contact with the hotter tank walls. The heat flux along the free surface has been considered to be null.

Figure 5.7 shows some snapshots of the numerical simulation. The temperature contours have been superposed on the fluid domain at different time instants.

Figures 5.6 and 5.7 show that the fluid does not heat uniformly because of the convection effect that is automatically captured by the Lagrangian technique here presented.

Falling of Three Objects in a Heated Tank Filled with Fluid

This 2D example is taken from [6]. Three solid objects with the same shape fall from the same height into a tank containing a fluid at rest. For the solid objects a

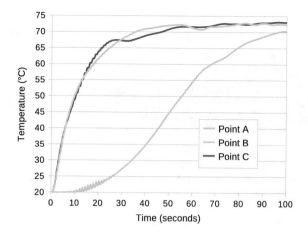

Fig. 5.6 2D sloshing of a fluid in a heated tank. Time evolution of the temperature at the points *A*, *B* and *C* of Fig. 5.5 [5]

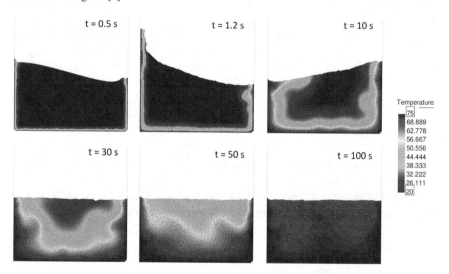

Fig. 5.7 2D sloshing of a fluid in a heated tank. Snapshots of fluid geometry at six different time instants. Colours indicate temperature contours [5]

hypoelastic model has been used. The geometry and the problem data, as well the initial thermal conditions, are shown in Fig. 5.8.

Once again, the material properties are assumed to be temperature independent. The fluid in the tank has initial temperature T = 340 K, while the solid bodies from left to right, have initial temperatures T = 180 K, T = 200 K, $T = 220$ K, respectively. The solid and the fluid domains have been discretized with a mesh composed of 9394 3-noded triangular elements. The simulation has been run for a total duration of 8 s using a time step increment $\Delta t = 0.0001$ s. The heat flux in the normal direction is assumed to be null for the boundaries in contact with the air or the walls.

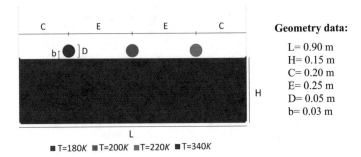

■ T=180K ■ T=200K ■ T=220K ■ T=340K

Fig. 5.8 Falling of three objects in a heated tank filled with fluid. Initial geometry, thermal conditions and material properties [6]

Fig. 5.9 Falling of three objects in a heated tank filled with fluid. Evolution of the temperature at the center of the three objects [6]

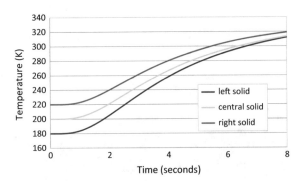

In the graph of Fig. 5.9 the evolution of the temperature at the central point of the three objects is plotted.

Figure 5.10 collects six representative snapshots of the numerical simulation with the temperature results plotted over the fluid and the solid domains.

5.4 Phase Change

For dealing with the phase change transformation, a term that takes into account the latent heat released or absorbed during the melting process needs to be considered [7]. Hence, the heat equation for phase change problems reads

$$\rho c \frac{DT}{Dt} - \frac{\partial}{\partial x_i}\left(k\frac{\partial T}{\partial x_i}\right) + Q + \rho \mathcal{L}_{PC}\frac{\partial f_L}{\partial t} = 0 \qquad (5.17)$$

where $\mathcal{L}_{PC}$ is the latent heat of the phase transformation and f_L is the liquid fraction function which is equal to one in the liquid phase and null in the solid one, as shown in Fig. 5.11.

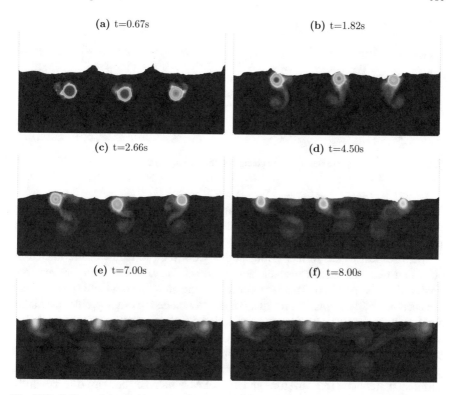

(a) t=0.67s   (b) t=1.82s

(c) t=2.66s   (d) t=4.50s

(e) t=7.00s   (f) t=8.00s

Fig. 5.10 Falling of three objects in a heated tank filled with fluid. Snapshots with temperature contours at different time steps [6]

Fig. 5.11 General liquid fraction function for phase change modeling

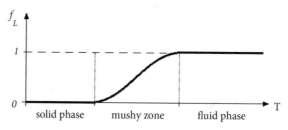

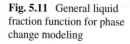

The phase change is modeled in the following way.

The solid elements transform into fluid elements if at least one of the two following conditions is verified:

1. *Due to temperature*: if the mean temperature of the solid elements that share the same node is greater than the melting temperature of the material. In this case, all these solid elements shift to the fluid domain, as illustrated in Fig. 5.12 for a 2D problem.
2. *Due to plastic deformation*: if a solid element has accumulated a plastic deformation greater than a pre-defined limit value.

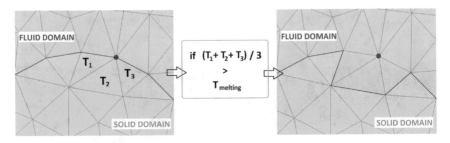

Fig. 5.12 Graphic representation of the change of phase algorithm

The transformation from fluid to solid or viceversa entails just a few of changes and small computational work. The main ones concern the change of the material properties and the remeshing procedure.

With respect to the former point, in this work the properties of the melted or solidified material change instantaneously taking the values of the other material analyzed in the problem. This represents a strong simplification and it may affect the accuracy of the numerical results. For a better modeling of the problem, a multi-fluid analysis should be performed considering different properties for the melted or solidified materials. Despite this more accurate strategy has not been implemented or tested, from the author point of view, it can be easily linked to the Unified formulation. This opinion is sustained by the fact that the proposed strategy is open to analyze different materials at the same time and the PFEM has already been used successfully in the past for solving multi-fluid problems [4].

Regarding remeshing, as already explained in the previous chapters, the mesh is regenerated just on the fluid domain. So when a solid element passes to the fluid domain, its nodes, except the ones that still belong to the contour of the remaining part of solid domain, become involved in the remeshing procedure, as the rest of the elements in the fluid domain. The remaining part of the solid domain is not remeshed but it just loses the part of its domain involved in the melting.

The above two changes are essentially the only ones required by the thermally coupled Unified formulation for dealing with phase change problems.

5.4.1   Numerical Example: Melting of an Ice Block

This 2D example is presented to show an application of the phase change strategy implemented. The problem was presented in [6]. An ice block at initial temperature $T = 270$ K is dropped into a tank containing water at rest at temperature $T = 340$ K. The initial geometry, the problem data and the initial thermal conditions are shown in Fig. 5.13.

Ice is treated as a hypoelastic solid until some of its elements reach the fusion temperature ($T = 273.15$ K). These elements pass to the fluid domain taking all its

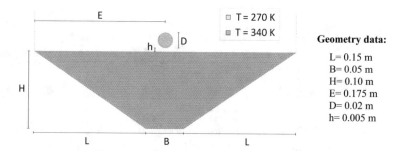

Fig. 5.13 Melting of an ice block. Geometry, material data and initial thermal conditions [6]

physical properties. For this analysis the following assumptions have been made: the mechanical and thermal properties of the water and the ice do not change with the temperature and the heat normal flux along the boundaries in contact with the air or the walls have been considered to be null.

In Fig. 5.14 snapshots of some representative instants of the analysis are shown. The temperature contours are plotted over the water and the ice.

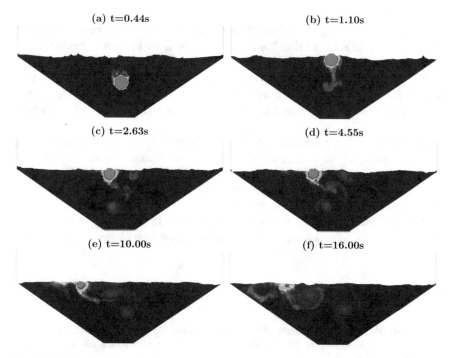

Fig. 5.14 Melting of an ice block. Snapshots with temperature contours at different time steps [6]

(a) t=2.81s   (b) t=4.70s   (c) t=9.77s

(d) t=12.74s   (e) t=14.69s   (f) t=15.47s

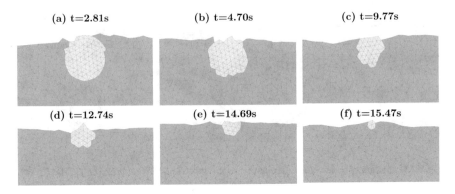

Fig. 5.15 Melting of an ice block. Zoom on the piece of ice at different time steps [6]

In Fig. 5.15 the detail of the melting of the ice piece is illustrated. The finite element mesh is drawn over the solid and the fluid domains.

5.5 Summary and Conclusions

The purpose of this chapter has been to show that the Unified PFEM formulation can be easily coupled with the heat transfer problem in order to solve coupled thermal mechanical problems.

The coupling has been ensured via a staggered scheme for which the mechanical and the thermal problems are solved within the same iteration loop. For the temperature field the same linear shape functions of the velocity and pressure fields have been used.

Several numerical examples have been presented with the objective of showing the applicability of the Unified coupled thermal–mechanical formulation for solid and fluid dynamics problems involving the temperature.

The algorithm for the phase change modeling has been explained and an explicative numerical example has been also given.

The numerical examples presented have shown the possibility of the Unified PFEM formulation for dealing with thermal–mechanical problems.

References

1. E. Oñate, R. Rossi, S.R. Idelsohn, and K. Butler. Melting and spread of polymers in fire with the particle finite element method. *International Journal of Numerical Methods in Engineering*, 81 (8):1046–1072, 2010.
2. E. Oñate, J. Marti, R. Rossi, and S.R. Idelsohn. Analysis of the melting, burning and flame spread of polymers with the particle finite element method. *Computer Assisted Methods in Engineering and Science*, 20:165–184, 2013.

3. P. Ryzhakov. *Doctoral thesis: Lagrangian FE Methods for Coupled Problems in Fluid Mechanics*. 2010.
4. S.R. Idelsohn, M.Mier-Torrecilla, and E. Oñate. Multi-fluid flows with the particle finite element method. *Computer methods in applied mechanics and engineering*, 198:2750–2767, 2009.
5. E. Oñate, A. Franci, and J.M. Carbonell. A particle finite element method (pfem) for coupled thermal analysis of quasi and fully incompressible flows and fluid-structure interaction problems. *Numerical Simulations of Coupled Problems in Engineering. S.R. Idelsohn (Ed.)*, 33:129–156, 2014.
6. E. Oñate, A. Franci, and J.M. Carbonell. A particle finite element method for analysis of industrial forming processes. *Computational Mechanics*, 54:85–107, 2014.
7. N. Dialami. *Doctoral thesis: Thermo-Mechanical Analysis of Welding Processes*. 2014.

Chapter 6
Industrial Application: PFEM Analysis Model of NPP Severe Accident

6.1 Introduction

In this chapter the analysis of an industrial problem solved using the Unified formulation is presented. The project required the modeling of the damages in a pressure vessel structure caused by the dropping of a volume of corium at high temperature.

In a nuclear power plant, the reactor pressure vessel (Fig. 6.1) is the structure that contains the reactor core, the nuclear reactor coolant and the core shroud.

The core of a nuclear reactor is the place where the nuclear reactions occur. The coolant is used to remove the heat from the nuclear reactor and the core shroud is a strainless structure used to direct the cooling water flow.

Corium is a heterogeneous material, also called Fuel Containing Material (FCM) or Lava-like Fuel Containing Material (LFCM). Its formation is the result of the combustion and melting of the reactor's components and the subsequent chemical and radioactive reactions. Consequently, the corium's composition depends on the type of the reactor and specifically on the materials of its components. Generally it is composed by nuclear fuel, fission products, control rods, structural materials from the affected parts of the reactor, products of their chemical reaction with air, water and steam, and, in case the reactor vessel is breached, molten concrete from the floor of the reactor room.

The temperature of corium depends on its internal heat generation dynamics, specifically on the chemical and radioactive reactions that may occur during the accident. Potentially, the corium can reach temperatures over 2800 C.

Corium accumulates at the bottom of the reactor vessel and, in case of adequate cooling, solidifies. Otherwise, the corium may melt through the reactor vessel and flow out (Fig. 6.2).

The seriousness of this kind of accident explains the high interest of this study.

The purpose of the analysis was to model with the PFEM the interaction between the corium and the pressure vessel structure. The scope was to evaluate the capability

© Springer International Publishing AG 2017
A. Franci, *Unified Lagrangian Formulation for Fluid and Solid Mechanics,
Fluid-Structure Interaction and Coupled Thermal Problems Using the PFEM*,
Springer Theses, DOI 10.1007/978-3-319-45662-1_6

Fig. 6.1 Pressure vessels for nuclear power plants. From [1]

Fig. 6.2 A blob of corium in
the Chernobyl Nuclear
Reactor. From [2]

of the method for simulating such a complex problem. For the project, the PFEM Unified formulation with thermal coupling presented in this work has been used.

The study is a multi-physics and highly non-linear problem and it involves many critical topics for a computational analysis, such as free-surface flow, fluid-structure interaction, plasticity with thermal softening and phase change.

Specifically, the analysis consisted of two models:

- Basic model. A sphere of corium falls from a certain height over a prismatic plate;
- Detailed model. A corium volume is placed all around a control rod housing located in the center of a steel shell.

The first model represents a general case and its geometry is not necessary related to a specific component of the pressure vessel. In Fig. 6.3 a graphic and simplified representation of the study is given.

Instead, in the second model the geometry is more complex and it reproduces a part of the bottom head of a reactor pressure vessel. In Fig. 6.4 a graphic representation of the area of study is shown.

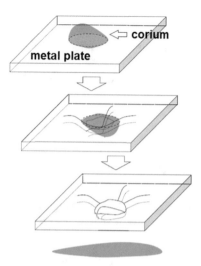

Fig. 6.3 Basic model. Graphic representation of the phenomena required to model

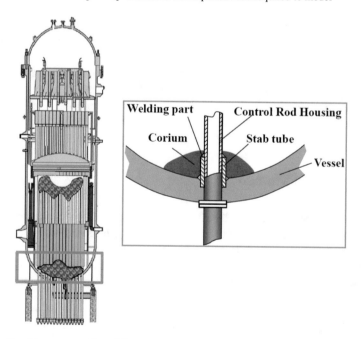

Fig. 6.4 Graphic representation of the accumulation of corium at the bottom of the pressure vessel (images provided by NSSMC)

For both models, the phenomena to model are:

1. Dropping and slumping of a high temperature volume of corium on the pressure vessel;
2. Melting, deforming and destruction of the structure due to the heating caused by the corium;
3. Slumping of corium through the vessel.

6.1.1 Assumptions Allowed by the Specification

Due to the complexity of the analysis the following assumptions were considered to be admissible in the specifications provided by the contractor for the analysis of the two models:

- The data analysis given in the specifications (geometric data, material properties, boundary conditions…) were open to changes after consultation with the contractor;
- Simplified constitutive models could be considered whenever necessary;
- The dimensions of the models could be scaled down in consideration to the speed of the analyses;
- The cooling water and its effects (boiling, flow etc.) are not considered in the study;
- The corium is assumed to be a highly viscous fluid at constant temperature;
- The corium could not mix or react with the surrounding media;
- The melting and solidification of steel are treated with a simple model.

6.2 Numerical Method

For simulating both models described in the previous section, the Unified formulation with thermal–mechanical coupling has been used. The multi-physics of the project represented a great opportunity to verify the efficiency of the formulation in its completeness. In fact, the analysis involves most of the technical aspects presented in this work. The Unified formulation allows the simulation of the interaction of free-surface fluids with solids with plasticity, as shown in Chaps. 2–4. Finally Chap. 5 showed that with just a small implementation work, also coupled thermal–mechanical and phase change problems can also be simulated.

From the practical point of view, in industrial projects the computational cost of the analysis has to be taken in serious consideration. Due to the complexity of the (3D) geometry, the duration of the physical phenomena and the required accuracy of the results, some industrial problems may lead to prohibitive computational times. Hence, in some cases it is necessary to accept a compromise between the accuracy and computational cost of the analysis. In order to calibrate at best the model, some preliminary studies of the simulations should be done. In this work, for example, this phase has been extremely useful for defining the adequate time step increment and the best FEM discretizations, deciding where a finer mesh was necessary and where not, and also for setting up some geometric and material values. For both models considered a preliminary study has been done and in the following some details of this phase are given.

Concerning the numerical method used for these simulations, the mechanical problem was solved with the Unified formulation using the VP-element for the solid with an hypoelastic-plastic model, a von Mises yield criterion and thermal softening. The phase change was modeled following the criteria presented in Sect. 5.4. In both models there is only melting of the solids and not solidification of the fluids. For simplicity, it was assumed that the solid elements that become fluid take instantaneously the same material properties of the fluid. As already explained, for a more realistic simulation, a multi-fluid analysis should be performed taking into account different properties for the melted material.

6.3 Basic Model

6.3.1 Problem Data

The basic model consists of a corium sphere at $T = 2000\,\mathrm{K}$ that falls from a height H over a prismatic plate at $T = 293\,\mathrm{K}$.

The geometry of the problem is illustrated in Fig. 6.5.

The boundary conditions of the plate are illustrated in Fig. 6.6.

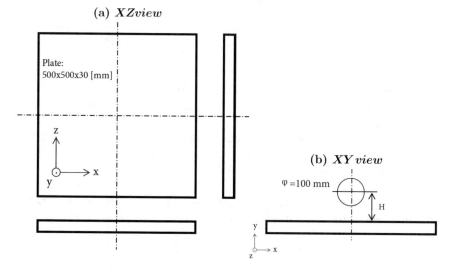

Fig. 6.5 Basic model: XZ and XY views of the initial geometry

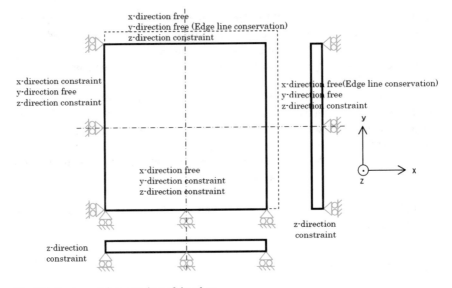

Fig. 6.6 Basic model: constraints of the plate

Table 6.1 Material data for the corium at T = 2000 K

ρ	k	c_v
kg/m³	W/(m · K)	J/(kg · K)
7865	22	500

Table 6.2 Material data for the plate at T = 293 K

ρ	E	ν	k	c_v	α	σ_Y	$T_{melting}$
kg/m³	GPa	–	W/(m · K)	J/(kg · K)	1/K	GPa	K
7850	196	0.33	16	500	0.000012	0.282	1693

The properties of the corium are the ones given in the specification for $T = 2000$ K and they are collected in Table 6.1. It has been assumed that the corium keeps the same temperature during the analysis.

Concerning the steel plate, the material data at $T = 293$ K are collected in Table 6.2. These material properties are comparable to those of the stainless steels and they can represent a Stainless Steel 309.

The thermal influence on the material properties has been taken into account by multiplying the Young modulus E, the heat conductivity k, the specific heat c_v and the yield stress σ_Y by the coefficients C_E, C_k, C_c and C_Y, respectively. Hence:

$$E(T) = C_E(T) \cdot E(T = 293 \text{ K}) \tag{6.1}$$

$$k(T) = C_k(T) \cdot k(T = 293 \text{ K}) \tag{6.2}$$

Table 6.3 Thermal coefficients for the plate

Temperature		Thermal coefficients			
[C]	[K]	C_E	C_k	C_c	C_Y
20	293	1	1	1	1
93	366	0.992	1.082	1.036	1
204	477	0.969	1.208	1.091	1
315	588	0.914	1.333	1.146	1
427	700	0.910	1.460	1.202	1
538	811	0.864	1.586	1.257	0.781
649	922	0.828	1.711	1.312	0.767
760	1033	0.900	1.837	1.367	0.636
871	1144	0.778	1.962	1.422	0.267
982	1255	0.733	2.087	1.477	0.146
1093	1366	0.733	2.087	1.477	0.010
1377	1650	0.733	2.087	1.477	0

$$c_v(T) = C_c(T) \cdot c_v(T = 293 \, \text{K}) \tag{6.3}$$

$$\sigma_Y(T) = C_Y(T) \cdot \sigma_Y(T = 293 \, \text{K}) \tag{6.4}$$

In Table 6.3 the values of the thermal coefficients are given for certain temperatures. For other temperatures, the coefficients are obtained via a linear interpolation.

6.3.2 Preliminary Study

In order to calibrate the model, the problem was studied in 2D first. The original purpose of this preliminary study was just to understand better the difficulties of the problem and to determine a range of suitable values for both the mesh size and the time step increment. However, this preliminary study allowed to identify that some parameters of the proposed model needed to be modified in order to simulate the desired phenomenon of heating and melting of the steel plate.

In fact, in the initial configuration proposed in the specifications, the sphere of corium was located at the height $H = 200$ mm. However, the preliminary studies showed that it would be not possible to simulate the desired problem if the corium would fall from that height. In fact, the inertial forces accumulated by the corium during its fall generate a huge splashing of the fluid over the plate. In order to allow the laying of the corium on the plate and the consequent heating of the structure, the initial height of the corium was reduced in the PFEM analyses. After some tests the height $H = 80$ mm was chosen for the simulation.

The preparatory tests also showed that with the value of corium viscosity proposed in the specifications ($\mu = 0.01$ Pa · s), the corium flows progressively towards the edges of the plate and finally a part of this falls down by the plate contour. In order to avoid this undesired phenomenon, a non-linear viscosity was used for the basic model. In particular, corium was modeled using the constitutive law proposed and successfully applied for the simulation of melted metals in [3, 4]. According to the mentioned publications, the viscosity of the corium was computed as:

$$\mu = \frac{\sigma_Y(T)}{\sqrt{3}\bar{\epsilon}_d} \tag{6.5}$$

where $\bar{\epsilon}_d$ is the deviatoric strain invariant and σ_Y is the yield stress.

6.3.3 Numerical Results

The simulation was run using a finite element mesh composed by 327,451 four-noded tetrahedral elements with 61,452 nodes; 27,936 nodes for the fluid and 33,516 for the solid. The fluid was discretized using an uniform mesh with an average size of 5 mm. On the other hand, the solid plate was not meshed uniformly. In its central part, for guaranteeing a good contact between the solid and the fluid, an unstructured mesh with the same average size of the fluid was used, while for the surrounding zone the average mesh size is 8 mm.

For the first phase, when the splashing occurs the time step increment $\Delta t = 0.002$ s was used. In Fig. 6.7 some snapshots referred to this phase are given.

In order to reduce the computational time of the analysis, for the other phases of the simulation the time step increment was increased to $\Delta t = 0.02$ s.

Fig. 6.7 Basic model. Snapshots of the initial splashing of the corium on the plate

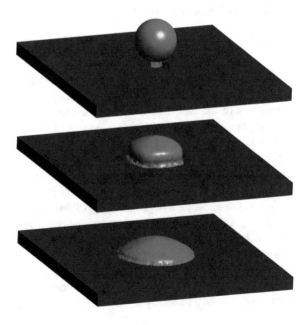

Fig. 6.8 Basic model. Snapshots of the melting of the plate and plot of temperature contours (blue and red colors correspond to 293 and 2000 K, respectively)

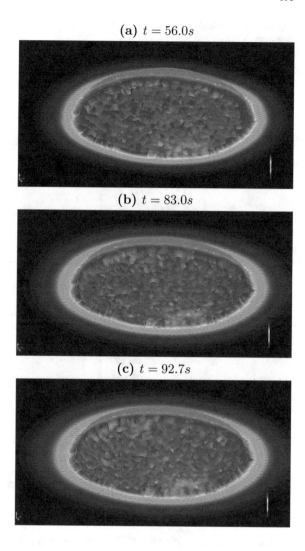

(a) $t = 56.0s$

(b) $t = 83.0s$

(c) $t = 92.7s$

After the splashing and spreading of the corium on the steel plate, the plate starts heating due to the contact with the hotter corium and progressively melts in its central part. In Fig. 6.8 a few snapshots of this phase are given.

The melting of the plate starts at $t = 30.3$ s. The time evolution of the accumulated melted volume of the solid material is illustrated in the graph of Fig. 6.9.

From the graph one can deduce that the law is non-linear. This behavior is in part a consequence of the hypothesis of constant temperature for the corium. For this reason the melted particles of the solid elements take immediately the temperature of the corium that is greater than the melting temperature of the solid. A more reliable law would be obtained by considering variable the temperature of the corium and by using a higher heat capacity for the solid elements.

Fig. 6.9 Basic model. Time evolution of the melted volume of the steel plate

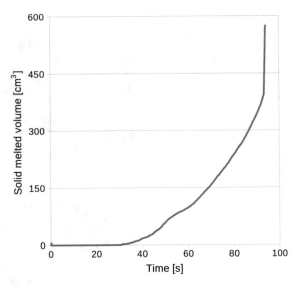

Fig. 6.10 Basic model. Time evolution of the temperature at the center of the lower side of the plate

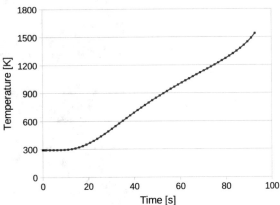

Another interesting point of the graph is represented by the increasing of the velocity of melting during the final instants of the simulation. The reason is that during this phase there is an increasing number of solid elements that become part of the fluid domain not due to the temperature criterion (see Fig. 5.12) but due to the plasticity effect. In fact, in that moment of the analysis a huge part of the width of the plate has already melted and the thin remaining layer of the plate has to sustain all the weight of the corium and the melted steel. In this situation, that zone of the plate plastifies and it undergoes huge plastic deformations. A proof of this can be deduced from the graph of Fig. 6.10.

(a) $t = 93.8s$

(b) $t = 94.1s$

(c) $t = 94.4s$

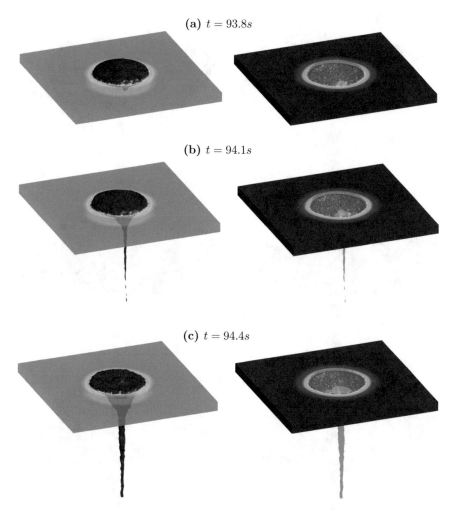

Fig. 6.11 Basic model. Snapshots of the piercing of the melted plate by the corium (I/II)

The evolution of the temperature in the plate bottom does not present that pronounced peak in the final part. This explains that the final increasing in the number of melted elements is not related directly with the effect of the temperature.

The collapsing of the plate occurs at $t = 93.8$ s. The hole created by the corium enlarges quickly due to the large deformations accumulated in the center of the plate.

This effect can be clearly seen in Figs. 6.11 and 6.12, where several snapshots of this phase of the simulation are given.

(a) $t = 94.5s$

(b) $t = 94.6s$

(c) $t = 94.8s$

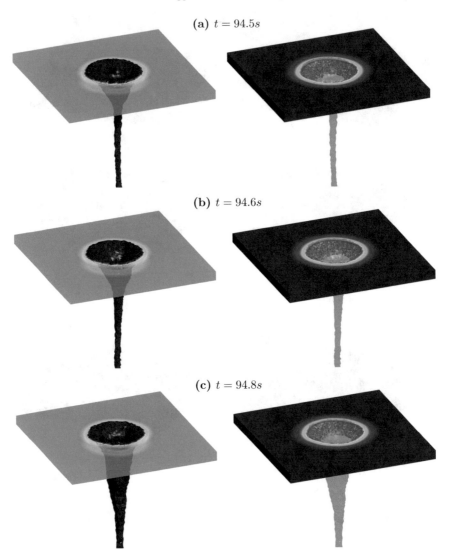

Fig. 6.12 Basic model. Snapshots of the piercing of the melted plate by the corium (II/II)

In Fig. 6.13 the failure of the plate is represented in its deformed configuration enlarged by a factor of 10^5. From this representation it can be appreciated better the localization of the strains in the central part of the plate.

In Fig. 6.14 the plate deformation at the time $t = 93$ s is illustrated in a 3D-view with the same enlargement factor (10^5).

In Fig. 6.15 the XZ-view of the steel plate at the end of the analysis is given.

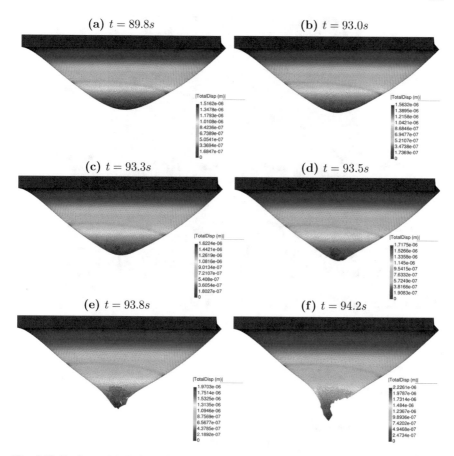

Fig. 6.13 Basic model. Deformed mesh of the plate during the last phase of the simulation (enlargement factor = 10^5)

Fig. 6.14 Basic model. 3D-view of the deformed mesh of the plate at t = 93.0 s (enlargement factor = 10^5)

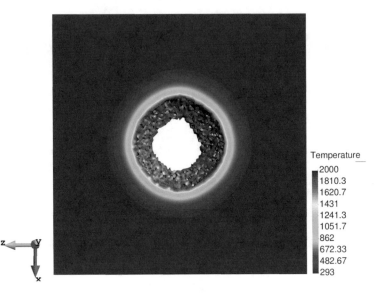

Fig. 6.15 Basic model. Final configuration of the steel plate

6.4   Detailed Model

6.4.1   Problem Data

The problem consists of a volume of corium at $T = 2000$ K that leans against a steel shell and surrounds a rod that is located in the middle of the steel structure. In Fig. 6.16 the initial geometry of the problem is illustrated. All the solid components of the problem have an initial temperature of $T = 293$ K. The objective of the analysis is to simulate the melting of the solid structure due to the heating caused by the corium.

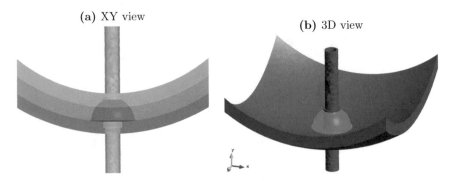

Fig. 6.16 Detailed model. 3D views of the initial geometry

The shell is constrained in the same manner as the plate of the basic model Fig. 6.6 while the rod is constrained at its top.

Concerning the plate and the corium, the material properties are the ones already introduced for the basic model (Tables 6.1, 6.2 and 6.3). As for the basic model, it has been assumed that the corium keeps the same temperature ($T = 2000$ K) for all the duration of the analysis.

6.4.2 Preliminary Study

In the detailed model the steel structure is not plane as for the basic model, so the corium can not spread over the structure and fall by its edges. For this reason, in this case, it was not necessary to use the non-linear viscosity of Eq. (6.5).

However, with the proposed value of the viscosity $\mu = 0.01$ Pa · s, the corium should have a water-type behavior. In fact, the ratio ρ/μ of the corium is very similar to that of water:

$$[\rho/\mu]_{corium} = 7865/0.01 = 0.7865 \cdot 10^6 \tag{6.6}$$

$$[\rho/\mu]_{water} = 1000/0.001 = 1.0 \cdot 10^6 \tag{6.7}$$

It is well known that the Reynolds number depends on this ratio. Hence, the dynamics of the problem involving corium should be similar to that of the same problem involving water. With the proposed value of viscosity, the corium initially splashes over the shell, then starts oscillating until a hydrostatic state is reached. This is in contradiction with the description of corium as a 'highly viscous fluid' given in the specifications. Furthermore, from the computational point of view, for modeling this kind of problem a small time step increment should be used and the discretization of the upper surface of the plate should be fine. The union of these two requirements would increase highly the computational cost of the analysis.

For all these reasons, a higher value of viscosity has been considered. More specifically, $\mu = 1$ Pa · s has been used.

6.4.3 Numerical Results

The finite element mesh used for the simulation consists of 396,984 four-noded tetrahedra (71,992 nodes). Both the fluid and the solid have been discretized using a non-structured mesh. In the central parts of the shell and the rod a finer mesh has been adopted. In particular, in those zones of the solid and in all the fluid domain the mesh has an average size of 3.8 mm.

As for the basic model, the time step increment has been set depending on the phase of the simulation. In the first phase, when the corium is spreading on the shell and around the rod, the time step increment $\Delta t = 0.002$ s was used. In Fig. 6.17, some snapshots of this phase are given. After 0.6 s the viscous fluid is almost at rest.

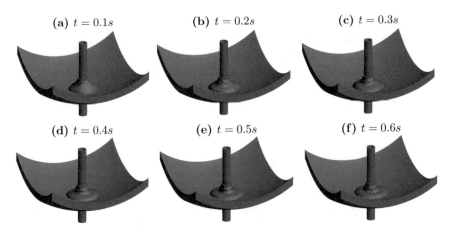

(a) $t = 0.1s$ (b) $t = 0.2s$ (c) $t = 0.3s$

(d) $t = 0.4s$ (e) $t = 0.5s$ (f) $t = 0.6s$

Fig. 6.17 Detailed model. Snapshots of the initial spreading of corium

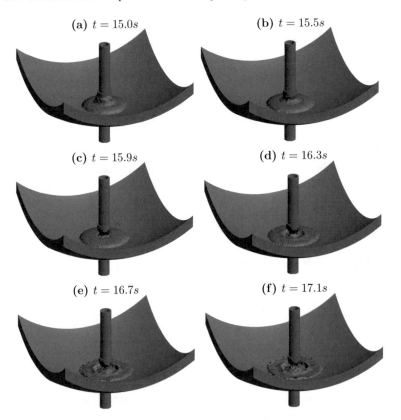

(a) $t = 15.0s$ (b) $t = 15.5s$

(c) $t = 15.9s$ (d) $t = 16.3s$

(e) $t = 16.7s$ (f) $t = 17.1s$

Fig. 6.18 Detailed model. Snapshots of the final phase of the simulation. A hole is created in the rod and the corium passes through it

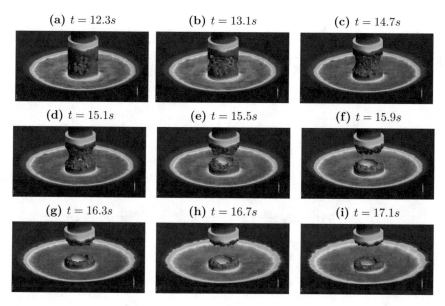

(a) $t = 12.3s$ **(b)** $t = 13.1s$ **(c)** $t = 14.7s$

(d) $t = 15.1s$ **(e)** $t = 15.5s$ **(f)** $t = 15.9s$

(g) $t = 16.3s$ **(h)** $t = 16.7s$ **(i)** $t = 17.1s$

Fig. 6.19 Detailed model. Snapshots of rod melting

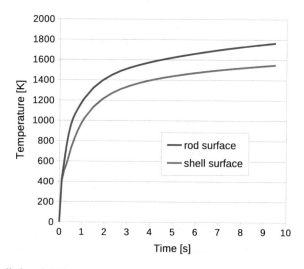

Fig. 6.20 Detailed model. Time evolution of the temperature on the rod and shell external surfaces

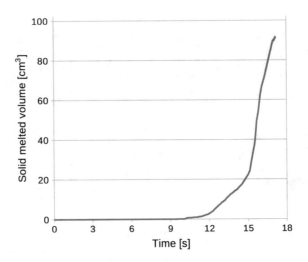

Fig. 6.21 Detailed model. Time evolution of the melted volume

For the rest of the simulation the time step increment was increased to $\Delta t = 0.005$ s.

In Fig. 6.18 some representative snapshots of the rest of the analysis are given.

The results show that the failure of the structure occurs in the rod. This can be clearly visualized in Fig. 6.19 where only the structure is represented: the central part of the rod is completely melted, while no parts of the shell have reached the melting temperature of the material.

In the graph of Fig. 6.20 the time evolution of the temperature on the external surfaces of the rod and the shell is plotted. This graph confirms that the rod is heating faster that the shell.

The melting of the rod starts at $t = 9.53$ s. The time evolution of the accumulated melted volume of the solid part of the domain is illustrated in the graph of Fig. 6.9 (Fig. 6.21).

Note that in this case the peak of melting velocity that occurs in the basic model (see Fig. 6.9) does not happen. That is because in this problem the phase change of the solid is essentially governed by the temperature. The plastic zones are local and they do not create any plastic mechanisms, as it occurs for the basic problem (see Fig. 6.13).

In Fig. 6.22 the deformation of the solid structure at $t = 15.0$ s is illustrated in a 3D-view with an enlargement factor of 10^5.

Fig. 6.22 Detailed model. 3D-view of the deformed configuration of the shell and the rod at $t = 15.0\,\mathrm{s}$ (enlargement factor $= 10^5$)

6.5 Summary and Conclusions

In this chapter the potentialities of the PFEM have been tested for the solution of two very complex problems related to the melting of two structures in two specific situations of NPP severe accident. The simulations involved many critical phenomena for a finite element analysis, such as free-surface flows, thermal fluid-structure interaction, and phase change.

The objectives of the study have been reached for both the basic and detailed models proposed. In particular, the phenomena required by the contractor have been modeled successfully.

The analyses were run by using a slightly different model than the proposed one. Preliminary tests indicated that with the geometry and viscosity of the corium proposed initially by the contractor, the fluid would splash and fall down without melting the adjacent structure. For this reason, with the agreement of the contractor, some input data were modified. In particular, the corium was modeled using a higher value of the viscosity and, for the basic model, a smaller initial height was considered.

Due to the complexity of the analysis, some approximations and simplifications were accepted. The purpose of most of these was to reduce the computational cost of the analyses.

Among these assumptions, the most invasive ones were the use of the constant temperature for the corium and the simplified modeling of the different material properties of the solid melted volume. In particular, the constant temperature for the corium may have accelerated the melting of the structure. A more accurate simulation would be obtained by considering the corium to have a variable temperature and a higher heat capacity.

References

1. www.thehindu.com/news/national/tamil-nadu/kudankulam-plant-not-to-draw-water-from-pechipaarai-dam-tamirabharani/article3443184.ece and http://www.world-nuclear-news.org/nn_nuclear_plans_forge_ahead_160609.htm.
2. http://islandbreath.blogspot.it/2013/07/fukushima-burns-on-and-on.htm).
3. O.C. Zienkiewicz, P.C. Jain, and E. Oñate. Flow of solids during forming and extrusion: Some aspects of numerical solutions. *International Journal of Solids and Structures*, 14:15–38, 1978.
4. O.C. Zienkiewicz, E. Oñate, and J.C. Heinrich. A general formulation for the coupled thermal flow of metals using finite elements. *International Journal for Numerical Methods in Engineering*, 17:1497–1514, 1981.

Chapter 7
Conclusions and Future Lines of Research

The aim of this chapter is to summarize all the thesis's work done highlighting its innovative points and main contributions. The thesis outcomes leave many open lines of research and future developments. The future lines of work are given in the last section of this chapter.

7.1 Contributions

The objective of this thesis was the derivation and the implementation in a C++ code of a unified formulation for fluid and solid mechanics, fluid-structure interaction and thermal coupled problems using the PFEM.

The finite element procedure was derived starting from the Velocity formulation presented in Chap. 2. In the same chapter, the mixed Velocity formulation was obtained exploiting the linearized form of the Velocity formulation. The mixed Velocity–Pressure scheme is based on a two-step Gauss–Seidel procedure where first the linear momentum equations are solved for the velocity increments and then the continuity equation is solved for the pressure in the updated configuration. Linear interpolation is used for both the velocity and the pressure fields. Both velocity-based finite element Lagrangian procedures were derived first for a general compressible material and then particularized for the hypoelastic model. The hypoelastic solid elements generated from the Velocity formulation and the mixed Velocity–Pressure formulations were called V and VP elements, respectively. The V and the VP elements were validated for several benchmark problems for elastoplastic compressible structures. It was shown that both elements are convergent for all the numerical examples analyzed.

© Springer International Publishing AG 2017
A. Franci, *Unified Lagrangian Formulation for Fluid and Solid Mechanics,*
Fluid-Structure Interaction and Coupled Thermal Problems Using the PFEM,
Springer Theses, DOI 10.1007/978-3-319-45662-1_7

In Chap. 3 the Unified Stabilized formulation for quasi-incompressible materials was derived. This numerical procedure essentially consists of the mixed Velocity–Pressure formulation derived in Chap. 3 where for the continuity equation the FIC stabilized form is used. The FIC stabilization was derived for the case of quasi-incompressible fluids and was extended also to hypoelastic quasi-incompressible solids. The stabilized element for quasi-incompressible solids was called VPS element. The solution schemes for solving quasi-incompressible Newtonian fluids and hypoelastic solids with the stabilized mixed Velocity–Pressure formulation were given and explained in detail. The entire Sect. 3.4 was devoted to the analysis of free surface fluids with the Unified stabilized formulation. First the Particle Finite Element Method (PFEM) was explained analyzing its advantages and disadvantages. An innovative strategy for modeling slip conditions in a Lagrangian way was also described. The excellent mass preservation properties of the PFEM-FIC stabilized formulation by solving a variety of 2D and 3D free surface flow problems involving surface waves, water splashing, violent impact of flows with containment walls and mixing of fluids. Also the conditioning of the scheme was studied and the effect of the bulk modulus on the numerical scheme highlighted. It was shown that using a scaled value for the bulk modulus, pseudo bulk modulus, in the linear momentum tangent matrix improves the conditioning and the global convergence of the linear system. A simple and efficient strategy for calibrating a priori the optimum value for the pseudo bulk modulus was also given. The strategy was successfully validated for two benchmark problems for free surface flow. In the last section of Chap. 3 the Unified stabilized formulation for fluids and solids at the incompressible limit was validated comparing the numerical results to both experimental tests and numerical results from other formulations. It was shown that the method is convergent to the expected solution. Specific attention was given to the study of the boundary conditions highlighting the beneficious effect of using slip conditions for inviscid fluids and coarse meshes.

Chapter 4 was devoted to the application of the Unified Stabilized to FSI problems. It was shown that this operation requires a small implementation effort. It is only required to assemble properly the global linear system and to detect the interface exploiting the capability of the PFEM for detecting the contours. It was shown that, depending on the specific need, one may choose any of the V, the VP and the VPS elements for modeling hypoelastic solids. The numerical solution given by the proposed scheme was validated with analytical solutions and numerical results of other formulations for free surface FSI problems.

In Chap. 5 the Unified formulation was used for solving coupled thermal-mechanical problem. The coupling was ensured via a staggered scheme and for the temperature field the same linear shape functions of the velocity and pressure fields were used. The numerical method was tested with several solid and fluid dynamics problems involving the temperature. The algorithm for the phase change modeling was explained and an explicative numerical example was given.

In Chap. 6 the Unified stabilized thermal coupled formulation was tested with an industrial problem. The analysis concerned the melting of two structures in two specific situations of Nuclear Power Plant (NPP) accident. The simulations involved many critical issues for a finite element analysis, such as free-surface flows, thermal fluid-structure interaction and phase change. Because of the complexity of the analysis, some approximations and simplifications were accepted by the contractor. The project ended successfully and all the phenomena required by the contractor in the specifications were modeled.

The innovative points and the contributions of this thesis can be summarized by the following list:

1. Derivation and implementation of the updated version of the FIC stabilization technique for quasi-incompressible fluids;
2. Study of the mass conservation in free surface fluid flows;
3. Extension of the FIC procedure to quasi-incompressible solids;
4. Analysis of the conditioning of the partitioned solution method for quasi-incompressible fluids;
5. Development of a procedure for improving the conditioning of the linear system;
6. Development of an innovative technique for treating the boundary nodes in the PFEM;
7. Slight modifications to the unified scheme for FSI problems proposed in [1];
8. Development of a simplified technique for phase change modeling.

7.2 Lines for Future Work

This section outlines the possible lines of research opened by this thesis. The following enumeration summarizes most of these

1. *New solid constitutive laws.* Although it has been shown that the hypoelastic model can be used for modeling many critical non-linear solid mechanics problems, it could very useful to extend the Unified formulation to other type of constitutive models (such as a hyperelastic model). In particular, it would be interesting to introduce a displacements-based model. This would be important also for showing the generality of the formulation;
2. *Improvement of the PFEM technology.* The PFEM technology implemented in this work allows the solving of one fluid only. Extension of the PFEM to multi-fluid would be of interest for many practical problems. For example, it has been remarked that in the viscous fluid problems the effect of the air has not been taken into account. This for some problems, i.e. casting process, can affect the truthfulness of the analyses. Previous works showed that the PFEM can deal with multi-fluid problems and this formulation does not constrain the implementation of a technology for the study of these problems. Another interesting improvement would be the modeling of the contact between solids and between a deformable solid and a rigid wall. This would extend the applicability of the proposed formulation;

3. *Generalization of the Lagrangian modeling of slip conditions*. In this work it has been presented a simple but efficient strategy for modeling slip boundary conditions in a Lagrangian way. The method has been applied to horizontal and vertical walls only. The technology has to be generalized for every type of walls. This could be done using Lagrangian multipliers or rotation matrices;

4. *Convergence analysis*. The convergence studies performed in this work were devoted to show qualitatively the convergence of the formulation more than evaluating and computing the convergence rate of the method. In order to do this, the L2 norm error should be studied. Its computation for free surface flow problems may be complex due to the high deformation of the domain and its discretization;

5. *Improvement and validation of the thermal coupled solver*. The coupling of the Unified formulation with the thermal problem was not the main objective of this thesis. The thermal coupling has been analyzed essentially with the purpose of showing the versatility and potential of the Unified PFEM formulation. For this reason, for the analysis of thermal coupled problems some simplifications have been assumed, above all for phase change modeling. The thermal and the mechanical solution were validated separately, however the work misses validation examples for coupled thermal-mechanical problem;

6. *Code optimization*. The computational time of the analysis is a not secondary issue. The C++ code implemented and used for this thesis may have not optimized routines that reduce the speed of the analyses. More work should be inverted in the future for reducing the computational cost of each function of the code;

7. *Parallelization*. The C++ code used in this work is fully sequential. In order to have a computational technology able to compete with commercial software, a parallel version of the code should be implemented.

Reference

1. S.R. Idelsohn, J. Marti, A. Limache, and E. Oñate. Unified lagrangian formulation for elastic solids and incompressible fluids: Applications to fluid-structure interaction problems via the pfem. *Computer Methods In Applied Mechanics And Engineering*, 197:1762–1776, 2008.

About the Author

Alessandro Franci was born in Arezzo, Italy, in 1986. He received the bachelor in Civil Engineering and the master degree in Structural Design from the Politecnico di Milano, Milan, Italy in 2008 and 2011, respectively. In September 2011 he started the Ph.D. in Structural Analysis at the Universitat Politècnica de Catalunya (UPC), Barcelona, Spain, under the guidance of Prof. Oñate and Dr. Carbonell. In 2014, he carried out a three-month predoctoral research stay at the Zienkiewicz Centre for Computational Engineering, Swansea, United Kingdom. In May 2015 he obtained the Ph.D. degree in Structural Analysis and his thesis was rated 'Excellent cum Laude' with international mention by the thesis defense committee. On December 2015, Franci won one of the PIONER awards assigned by the Catalan association CERCA. On February 2016, the Spanish Society of Numerical Methods in Engineering (SEMNI) awarded a SEMNI prize ex aequo to Franci for the best Ph.D. thesis in numerical methods in Spain for the year 2015. During the Ph.D. period, Franci has been the main developer of a couple of projects realized by the International Center for Numerical Methods in Engineering (CIMNE) for the Japanese company Nippon Steel and Sumimoto Metal Corporation (NSSMC) and devoted to the numerical simulation of a nuclear core melt scenario, one of the most severe accidents in a nuclear power plant. He has been teaching Structures and Material Resistance at the UPC since 2015. Currently, Franci is a post-doctoral researcher at CIMNE. His research is mainly focused on fluid dynamics and fluid-structure interaction problems accounting for thermal effects and on the application of the Particle Finite Element Method (PFEM) to industrial problems.

© Springer International Publishing AG 2017
A. Franci, *Unified Lagrangian Formulation for Fluid and Solid Mechanics,
Fluid-Structure Interaction and Coupled Thermal Problems Using the PFEM*,
Springer Theses, DOI 10.1007/978-3-319-45662-1

Printed in the United States
By Bookmasters